Ceramic transfer printing

Kevin Petrie

A & C Black • London

The American Ceramic Society • Ohio

Over the years I have received great advice and support from a number of individuals for which I am very grateful, so this book is dedicated to my mentors: Peter Searle, Julian Bray, John Houston, David Hamilton, Paul Van der Lem, Peter Davies, Brian Thompson and Allen Doherty.

In memory of my mother, Sheila Petrie, who was always a great support and in many ways still is.

First published in Great Britain in 2011
A & C Black Publishers Limited
36 Soho Square
London W1D 3QY
www.acblack.com

ISBN: 978-1-408-11328-8

Published simultaneously in the USA by
The American Ceramic Society
600 N. Cleveland Ave., Suite 210
Westerville, Ohio 43082, USA
http://ceramicartsdaily.org

ISBN: 978-1-57498-310-4

Typeset in 10 on 13pt Rotis Semi Sans
Book design by Susan McIntyre
Cover design by James Watson
Printed and bound in China.

This book is produced using paper that is made from wood grown in managed, sustainable forests. It is natural, renewable and recyclable. The logging and manufacturing processes conform to the environmental regulations of the country of origin.

COVER IMAGES, FRONT: (TOP) *A Dinner Service*, with 14 place settings, Charlotte Hodes, 2008. Hand-cut digital and screenprinted enamel transfers onto white china. *Photo: Peter Abrahams. Private collection.*
(BOTTOM) *Distorted Traditions*, Megan Randall, 2009. Approx. ht: 5cm (2in), tissue-printed transfers on porcelain. *Photo: courtesy of the artist.*

COVER IMAGES, BACK: (TOP) *Shopping Trolley Cake Stand*, Alice Mara. Ht: 34cm (13½in); small plate dia: 17cm (6¾in), medium plate dia: 20cm (8in), large plate dia: 27cm (10¾in), digitally printed bone china. *Photo: courtesy of the artist.*
(BOTTOM) *Head*, Kevin Petrie, c. 1994. Ht: approx. 35cm (13¾in), under-glaze tissue print on ceramic. *Photos: (left) courtesy of the artist, and (right) David Williams.*

FRONTISPIECE: *Floating*, Charlotte Hodes, 2006. Coloured slips, sprigs, digital and screenprinted enamel and gold transfers on earthenware, 41 x 35cm (16 x 13¾in). *Photo: Peter Abrahams, courtesy of Marlborough Fine Art.*

Contents

Acknowledgements

I would like to give my sincere thanks to all those who have freely offered their time to help me with this book. I am especially indebted to all the artists, museums and companies that have sent images and information.

In particular, I am grateful to the Victoria and Albert Museum, the Wedgwood Museum and the Potteries Museum, who have taken the first images of some pieces in their collections especially for this book. I spent an inspirational day at Royal Crown Derby and thank Louise Adams and all the staff for allowing me an insight into their fascinating work with transfers.

Many of the artists and researchers in this book kindly sent detailed reflections on their work, ideas and approaches, which in several cases I was able to use verbatim. This made my job more curatorial and editorial in some areas of the book, and importantly I feel it offers the artists' own perspectives on their work. In particular, Steve Brown, Kathryn Wightman, Elizabeth Turrel, David Sully, Tom Sowden, Shelley James, Rudi Bastiaans, Erik Kok and Charlotte Hodes all made huge contributions, for which I am very grateful. Also I would like to thank the Centre for Fine Print Research, University of the West of England for allowing me to include aspects of their work.

I have incorporated David Fortune's useful notes on firing schedules and the use of Daler Rowney products, and would like to acknowledge and thank him for his great help. Paul Scott's work in the area of ceramics and print remains an inspiration, and I have had valued discussions with him over the years. The University of Sunderland and my colleagues and students there have offered great encouragement and support, for which I am grateful. I would especially like to thank research student Megan Randall for taking the 'process' photographs in our ceramics and glass print studio at Sunderland; also Jeffrey Sarmiento for his 'tech notes' on the use of PhotoShop to develop positives.

Thanks also to all the team at A&C Black for turning my manuscript into the book that you are holding. In particular, I am very grateful to my commissioning editor, Alison Stace, who invited me to write this book, for her encouragement and sensitive editorial advice. Alison attended my first ceramic-transfer printing workshop, and it has been a pleasure to work with her again many years later.

Mandrake – Double Tuber, Steve Brown, 2009. Porcelain with underglaze, flexible in-mould transfer printing, 115 x 90 x 20cm (45¼ x 35½ x 7¾in). *Photo: courtesy of the artist.*

Introduction

Since the mid-18th century, ceramic transfer printing has enabled the application of a diverse range of printed aesthetic effects, in the form of pictures, pattern and text, to the surface of ceramics. As the name suggests, transfer printing enables the conveying of an image from one surface to another. 'Decal', an alternative term for 'transfer' often used in the USA, is derived from the French word *decalquer*, meaning to trace or copy. In simple terms, transfer printing involves printed imagery being transferred from paper to objects such as glass and ceramics. These prints are then fused to the surface of the objects through the action of heat. This requires the use of pigment that can withstand high temperatures. For clarity, the term 'transfer' is used throughout this book.

Printing is one of the prime methods of decorating ceramics. The advantage over hand-painting is that a print can be accurately and rapidly reproduced, making it especially attractive for industrial production. The individual artist or designer is less likely to be concerned with high-volume production, but might well be attracted to the diversity of aesthetic that print can offer. Hand-painting tends to look hand-painted, whereas a print can take many forms, from the painterly to the photographic.

There are many ways to print onto ceramics, and transfer printing forms a significant strand of the range of printing potential. Screenprinting is perhaps the most versatile print method available at present for printing onto ceramics, and so a large proportion of this book is devoted to it. Of course, direct printing also has great potential, but the advantage of using a transfer is that finer and more accurately registered imagery can be printed onto paper and then applied to the ware (registration is the accurate 'fit' of colours in printing). This is simply because it is much easier to print fine imagery onto paper than onto a piece of ceramic. In addition, unlike direct printing, a transfer is capable of decorating complex multi-curvature objects. Transfers also have the advantage that they can be cut up and freely collaged onto ceramics, creating a range of meanings, whereas a direct print is perhaps more limited.

Other methods besides screenprinting are considered in this book, as well as other surfaces that transfers can be applied to, such as enamels and glass. It is also worth pointing out that I mention some methods in Chapters 2 and 3 that I do not deal with later on in the book – lithography being a prime example. This is because I believe that screenprinting offers similar effects to lithography and is more accessible to the reader. However, I felt that it was important to discuss most transfer methods at least to some extent, as they might well have interesting applications in the future. I also hope that the historical context will in its own right be of interest to some and may offer inspiration to others in the way they approach the making of their art works.

A Dinner Service, with 14 place settings, Charlotte Hodes, 2008. Hand-cut digital and screenprinted enamel transfers onto white china. *Photo: Peter Abrahams. Private Collection.*

CASE STUDY: Royal Crown Derby – transfer printing in industry

The Royal Crown Derby Porcelain Company manufactures the highest-quality English fine bone china at its factory in Derby, England. Royal Crown Derby is one of the last-remaining ceramics factories in England, and a pre-eminent example of how screenprinted transfers are used for decoration. In my view, what is special about Royal Crown Derby ceramics is that they exploit the qualities of screenprinting as opposed to using it to reproduce designs originally made in other media, such as hand-painting.

This case study offers a glimpse of stages of production at this fascinating factory. Although the factory is operating at the highest level, the principles and methods they use are largely the same as those the studio artist will use, especially in terms of screenprinting, and so as a study forms a useful general overview of the processes involved.

ABOVE: The transfers are printed flat, but it is important that this decoration will fit the final form. Here the ceramic gorilla is being 'fitted' to provide a 2D plan, rather like a skin, to aid the design of the transfers, which are printed flat and then wrapped around the form. The yellow is covercoated flat colour used for test fits.

A peregrine falcon with the original painted design, Tien Manh Dinh, 2006. Ht: 20cm (8in).

RIGHT: From the original designs a model is made. Once this is approved, plaster moulds are made and the pieces cast in porcelain slip. During firing the clay shrinks by various degrees. So here David Harris is carefully measuring fired samples of an osprey in order to establish an average size so that the transfer-printed decoration can be designed to fit.

FAR RIGHT: Printer Martin Warren is seen here printing gold ink. This contains real gold and so has to be carefully managed – even cleaning rags are recycled to remove the precious material.

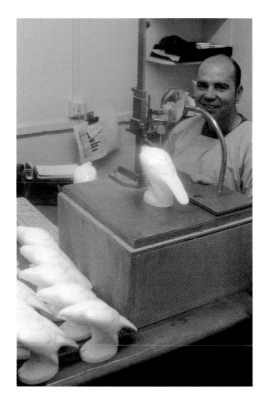

RIGHT: Here we see two in-glaze transfers for a mountain bluebird. Fired on examples can be seen on the ceramic bird. This creates a background for a second transfer of on-glaze and gold, which will now be applied. This can be seen in the top left of the picture.

FAR RIGHT: Louise Boyland applying transfers to a piece designed by the artist Ken Eastman.

Photos: Kevin Petrie

A personal perspective on transfer printing

In many ways this book draws heavily on my personal history and experience. In 1988 I started art school in the North Nottinghamshire town of Mansfield, where I first learnt printmaking. From here I went to the BA (Hons) Illustration course at what is now the University of Westminster, Harrow. This was a very broad-based course, and at one point I was told about ceramic transfer printing and found the materials, rarely used, in a cupboard. I knew how to screenprint and with some advice I taught myself transfer printing. I went on to devote my entire degree show to printed ceramics.

I was then accepted onto an MA course in Ceramics and Glass at the Royal College of Art in London. Here I learnt more about printing and explored early methods of transfer printing using potter's tissue and etching. I also printed a set of artist's plates as a gift for the retirement of John Hedgecoe, Professor of Photography, designed by his colleagues at the RCA. Having to print a diverse range of imagery was a great learning experience. It was around this time that I was first employed as a visiting lecturer, back at Harrow, to teach transfer printing. Students in these classes included Alice Mara, whose work is featured later, and the future commissioning editor of this book, Alison Stace.

Having left the Royal College, I heard about a position at the University of the West of England for a Ph.D. student to work on the development of water-based ceramic transfer printing. This took me to another city, Bristol. Here I helped develop a system within what is now the Centre for Fine Print Research with another Ph.D. student, Alison Logan, in the Faculty of Applied Sciences.

I completed my Ph.D. at the turn of the century, and soon afterwards came to work at the University of Sunderland as Lecturer in Ceramics, where I am now privileged to hold the position of Professor of Glass and Ceramics and lead the academic team in those subjects. During the decade that I have worked at Sunderland, I have largely explored the potential of glass and print, and my book of the same name forms the cornerstone of that work. The invitation to write this current book has allowed me to reflect back on my work with transfers, to revisit my Ph.D. and, importantly, to disseminate some of the exciting creativity and research taking place around the subject of ceramic transfer printing today.

It is important to say that each chapter of this book might be developed into a book in its own right. Therefore it is inevitable that some sections only offer a glimpse of what is possible. I hope therefore that readers will see this book as an introduction that they can use to help them to further explore their own individual approaches.

Policeman Flower Plate,
Kevin Petrie, (detail), *c.*1993.
Transfer on bone china.
Made when I was a BA
(Hons) Illustration student.
Photo: Kevin Petrie.

What is a ceramic transfer?

1

Before proceeding it is useful to explain some key principles.

Although there are a number of variations, the 'classic' transfer is known as the 'waterslide' transfer or decal. The most common type uses solvent-based materials but, as mentioned before, a water-based option is available. Differences between the two are discussed below as well as later in the book.

To produce standard solvent-based transfers, enamels are mixed into an appropriate printing medium and screenprinted onto a gum-coated transfer paper. Over this a layer called 'covercoat' is screenprinted. When thoroughly dry the transfer is soaked in water. This dissolves the gum layer and releases the covercoat and adjoined image from the backing paper. Transfers are applied to the glazed ceramic (or glass or enamel) face up and smoothed down with a cloth or rubber kidney. The residual gum from the transfer paper helps to adhere the transfer to the surface as well as lubricating it during positioning. During the firing the covercoat layer burns away cleanly, as does the printing medium, leaving the enamel image bonded permanently onto the ceramic surface.

The components of a waterslide transfer

LEFT: *Nostalgia*, Matt Smith, 2008. Ht: 31cm (12¼in). Thrown white earthenware with cast additions, under-glaze colours, custom and vintage transfers. *Photo: Gill Orsman.*

Let's take a journey through the components that make up a transfer, starting first with a solvent-based example. A transfer is essentially a printed image on paper. Imagine a cross section through the transfer with the print on the top. At the bottom is an absorbent base paper. This 'transfer' or 'decal' paper is pre-coated before you buy it with a layer of water-soluble gum. On top of this, there is the printed image. This is a combination of ceramic pigments, probably enamels, mixed with a printing medium that has been screenprinted onto the transfer paper. There might be several colours; each will have been printed after the preceding colour has dried.

Cross-section of a solvent-based ceramic transfer. *Diagram: courtesy of Chartwell Illustrators.*

Screenprint (enamels and solvent-based printing medium)

Covercoat layer

Water-soluble gum

Base paper

15

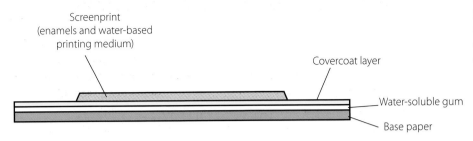

Screenprint
(enamels and water-based
printing medium)

Covercoat layer

Water-soluble gum

Base paper

Cross section of a water-based ceramic transfer on U-WET paper. *Diagram: Courtesy of Chartwell Illustrators.*

Once the entire image is complete, the final layer of the transfer is applied. This is a thermoplastic layer of 'covercoat', screenprinted, and slightly overlapping the image area. If you buy 'open stock' transfers or get them made for you as a 'custom run', there might be a sheet of interleaving paper, rather like greaseproof paper, which prevents the transfers from sticking together if stacked. Open stock refers to transfer-printed designs which can be bought off the shelf to decorate ceramics. Custom runs are transfers printed to a customer's specific design, and tend to be produced in quite high numbers.

A solvent-based transfer will be the most common type if you are buying 'open stock' or 'custom run' transfers. A water-based transfer, which you might choose to make yourself, is a little different from the solvent-based in that it requires a different type of transfer paper called U-WET. This paper was developed and patented by the University of the West of England after my Ph.D. there.

Imagine the cross section again. From the bottom we still have the absorbent base paper, and then the gum layer, but on top of this is a layer rather like covercoat but which covers the entire sheet of paper, not just the image. On top of this 'raft' of covercoat is the print, which again might be in several colours. There is no covercoat on top of this. The layer that the print 'sits' on serves the same purpose of conventional covercoat in that that it holds the image together during transferral. It is a different material and is less flexible than standard covercoat, but it does have the advantage of burning cleanly through the print during firing. Standard covercoat would not have this property if it were to be printed underneath.

CASE STUDY: Laura Straßer – evolving decoration

Laura Straßer is a German designer, located in Weimar and Leipzig in Germany. She was always fascinated by cobalt-blue patterns on Chinese ceramics and their European equivalents. In the history of blue and white decoration each new pattern is often an evolution of a former pattern, something she acknowledges in her design 'Mosaique'. Traditional Chinese decoration is utilised but in an almost insensitive manner. The embellishment has been cut up and put back together like a puzzle or mosaic, forming an ironic allusion to what Laura calls the 'uncaring and ignorant' decoration that exists today – and perhaps existed in earlier centuries too!

Laura Straßer, dragon design for a transfer for the 'Mosaique' range, 2008.
Photo: Courtesy of the artist.

2

A historical overview

Since the mid-18th century transfer printing has enabled the application of a diverse range of printed aesthetic effects to the surface of ceramics. This chapter traces key developments, and end notes point to further reading.

Transfer printing onto ceramics evolved from techniques used to decorate japanned and enamelled objects such as snuffboxes.[1] Although there is some conjecture over the exact inventor, it is thought that in the 1750s three men made important advances: John Brooks in printing enamels, Robert Hancock in printing porcelain, and John Sadler in printing earthenware and tiles.[2]

Basic principles of early transfer printing

Early methods of transfer printing primarily used engraved (or etched) metal plates. Etching and engraving are 'intaglio' printing techniques. Intaglio is the term used to describe a form of printing whereby the design is sunk below a surface, in this case a metal plate, and the cavities filled with ink. The surface of the plate is cleaned, leaving the ink contained in the hollows. The inked metal plate is then passed through a press with a piece of damp paper to give the printed impression. With engraving, the image is formed by incising grooves into the surface of the plate with a range of sharp chisel-like tools or punches. Etching differs from engraving in that instead of the line being cut or punched it is bitten into the metal with acid.

Early transfer printing can be divided into two stages: printing and transferring. In the case of intaglio transfer printing the ceramic colour, consisting of printing oil, metal oxides and fluxes, would be 'pushed' into all the intaglio areas. Excess ink was scraped from the surface of the plate, which was then 'bossed' with a leather-faced pad or similar to remove any remaining film of colour.

Transferral of the image: tissue and bat printing

'Italian Verandah' dish, Davenport, *c.*1835–40. Multi-colour single-plate print. *Photo: courtesy of The Potteries Museum & Art Gallery, Stoke-on-Trent.*

Historically there were two basic methods of transferring the image from printing plate to ceramic: 'tissue' printing and 'bat' printing. Bat prints were applied to glazed surfaces. The plates were charged with linseed oil, and a gelatin slab, or 'bat', was used to transfer the oil to the ceramic.[3] The bat was pressed onto the plate and then removed, drawing oil from the intaglio lines. It was then placed image side down onto the ware, leaving the printed image on the surface. The ware was then dusted with

Engraved plate. *Photo: William Blake, courtesy of the Potteries Museum & Art Gallery, Stoke-on-Trent.*

finely ground ceramic colour, which adhered to the oil. During firing the oil burnt away, leaving ceramic colour bonded permanently with the glaze.

Tissue printing involves transferral of the print from plate to ceramic via strong thin paper called 'pottery printing tissue' or 'potter's tissue'. The print is usually under the glaze. The image was first printed onto the tissue, which was then applied to the surface of the ceramic, image side down. Pressure was applied to the back of the tissue to offset the image onto the ware, with a little soft soap acting as lubrication. The print was then fired on in a kiln (see chapter 10).

The original bat printing process of the late 18th century was developed and mechanised in the 1950s with the invention of the 'Murray Curvex printing machine'.[4] A shaped pad of gelatin or rubber would pick up ink from engraved metal plates and transfer it onto ceramics. The gelatin pad was replaced by the 1980s with silicon.[5] The technologies used in silicon-pad printing and direct screenprinting have also been combined to produce offset screenprinting machines. In this process the image is printed by screen onto a flat transfer plate. It is then lifted from the plate and applied to ware via a silicon pad.

Direct screenprinting and offset screen-printing negated the need for engraved plates, which had several disadvantages, including weight of colour and an inability to produce solid blocks or bands of colour.[6] Having said that, engraved designs are still used in some sectors of the industry, especially in producing traditional designs such as the Willow Pattern.

Heat-release transfers allowed the automated application of transfers with specially formulated inks and covercoats printed onto a wax-coated paper.[7] The advantages of this process stem from savings in labour costs, as no soaking, squeegeeing or drying is required.[8]

Development of screenprinting

Screenprinting was used in the United States in the early 20th century for the production of banners, posters, advertisements and point-of-sale displays. The earliest existing example is in the Philadelphia Museum of Art, datable to c.1916 due to a 'patent pending' notice on the prints. The process described involves use of 'an open mesh screen, consisting of copper or brass-gauze, bolting cloth or other suitable material, tightly stretched on a rigid wooden or metal frame'.[9] The coloured image is then produced by successively blocking out the areas on the screen not to be printed. This is achieved by placing an original drawing or 'master pattern' underneath the screen and 'blocking out' all areas which are to remain unprinted by painting the screen with 'shellac, varnish, glue, asphalt, or any other suitable medium'.

The authors state that the mesh of the screen can also be masked by cutting stencils 'from paper, tissue, metal or any material best suited for the work in hand'. The colour is applied by placing the substrate to be printed underneath the screen and marking its

The potter's printer at work, 1919–21. *Photo: courtesy of Stoke-on-Trent Museums.*

position so that exact registration can be maintained. The colour (or ink) is then poured onto the screen surface and forced through the open areas of mesh (those that are not blocked out) by passing a flexible scraper, commonly known as a squeegee, or roller over the surface of the screen. The colour applied is thus forced through the open mesh of the screen and will adhere to the surface of the card and cover the parts desired. The screen is then cleaned and the successive colours printed in the same manner. The process detailed in this patent is a basic description of screenprinting, as it is known today.

An earlier British patent exists in the name of Samuel Simon, an artist and designer from Manchester, which was accepted in July 1907.[10] Simon's patent describes a process similar to that in the later 1918 patent, but in less detail.

Since the rather crude process described in the first patents for screenprinting, the technique has reached high levels of sophistication in recent years. Modern screens are made from synthetic materials, which are durable and improve print quality.[11] Stencils are commonly produced by 'photo-mechanical' methods, which allow for intricate detail. Printing is still carried out manually by hand-pulling the squeegee, but semi- and fully automatic machines are often used in the industry. Modern screenprinting can produce a wide range of printed products including self-adhesive labels and stickers, flexible circuits, T-shirts and printed textiles, wallpaper and floor coverings.

As well as becoming an important commercial medium, screenprinting has become an important means of artistic expression. According to R. and D. Williams, the first known artistic screenprints were made by the American artist Guy Maccoy in 1932.[12] In the 1930s several artists experimented with screenprinting but found it difficult to persuade galleries to exhibit their efforts. It was felt that the main difficulty was the name. Therefore a new name, 'serigraph', was devised by Carl Zigrosser, at the time Director of the Weyhe Gallery in New York City. The term 'serigraph', derived from the Latin words 'seri' meaning silk and 'graph' meaning write, allowed the creative screenprint to be distinguished from its commercial counterpart. Apparently, this ploy worked and the serigraph began to be accepted by galleries.

The outbreak of World War II curtailed many of the promising developments in the use of screenprinting as a creative tool. Artistic screenprinting was thus largely forgotten – perhaps due to the advance of abstract expressionism, a movement that in general did not favour printmaking – until its revival in the 1960s by Pop artists such as Andy Warhol.

Screenprinted ceramic transfers

Screenprinted ceramic transfers were first developed by the English company Johnson Matthey in the 1930s and were initially used for the labelling of bottles.[13] By the 1940s screenprinted transfers had been introduced to the ceramics industry. The process was further developed by the introduction in around 1954/5, again by Johnson Matthey, of the 'waterslide covercoat transfer', which could be described as the standard transfer of today.[14] The principle of the waterslide transfer is also used in the more recent developments in digital transfers (see Chapter 9).

Dr Cathy Treadaway, Research Fellow at the University of Wales Institute Cardiff, was one of the early pioneers of computer designed transfers for tableware. In the 1980s, her research was sponsored by Royal Doulton and two computer companies: IO Research and RM computers. Cathy devised methods of working with software to develop pattern fittings and designs, which were printed out on paper. She then used these to develop designs for prototype tableware sets and ceramic tiles.

Paul Scott – *Ceramics and Print*

Artist, writer, curator and educator Paul Scott is a seminal figure in the history of printing onto ceramics. In the 1990s he was central in drawing together the disparate information available about printing onto ceramics and presenting it in formats useful for artists. This was focused around his book *Ceramics and Print* (also published by A&C Black) and the exhibition Hot off the Press. *Ceramics and Print* was written because in the early 1990s there was little or no information on the creative application of print and ceramic processes. Scott spent a great deal of time working out how to use basic print processes with ceramic materials and 'decided that no one else should have to waste this kind of time again'. Published in 1994, *Ceramics and Print* is credited as an important catalyst in the phenomenal growth of printmaking within studio ceramics in the past decade or so.

Scott's studio work has also been influential. In particular he has drawn upon and subverted the convention of traditional blue and white pottery to comment on issues connected with contemporary landscape – for example, nuclear power, and the outbreak of foot-and-mouth disease that had a great impact on rural communities in the UK a few years ago. Recent work has been linked to a Ph.D. dealing with landscape and representations of landscape on ceramics.

Scott's Cumbrian Blue(s), Vindsäter Vignette, *Workers*, Paul Scott, 2009. ht: 15cm (6in), in-glaze transfers on tin-glazed hand-built forms, *Photo: courtesy of the artist.*

Mugs, Kevin Petrie, *c.*1999. Water-based transfers on bone china. *Photo: David Williams.*

Water-based ceramic transfer printing

It was around the same time that *Ceramics and Print* published in 1995, that I first started researching the potential of ceramic transfer printing in a formal way through a Ph.D. at the Bristol Print Centre – later to become the Centre for Fine Print Research at the University of the West of England, Bristol.

At the time the system used both by industry and artists contained solvent-based materials, in the form of printing mediums, covercoats and cleaning agents. Industrial solvent-based printmaking in general has become increasingly less viable in recent years due to health, safety and environmental legislation, which has limited the permissible levels of solvent in any given substance. This has led the industry to explore alternative systems, such as ultraviolet curing inks and water-based inks.[15]

The research project stemmed from the hypothesis that ideas and expertise diffused within the Bristol Print Centre could be channelled into evolving a workable reduced-solvent ceramic transfer printing system. The system we developed uses TW Flat Clear Base as a printing medium in conjunction with U-WET transfer paper, which was also developed at UWE. A parallel Ph.D. research project was also initiated by the Faculty of Applied Sciences.[16] This examined issues associated with the chemistry of water-based transfer printing, including the testing of water-based transfer printing in dishwasher and lead-release trials.

The work I undertook built on that of Dave Fortune (as well as that of Steve Hoskins and Richard Anderton), who has worked at UWE for many years managing the water-based screenprinting department. Dave disseminates his work through workshops at schools, colleges and universities, and also produces affordable screenprinting equipment for education and home printmakers. He has since developed an alternative ceramic transfer process using Daler Rowney System 3 additives; this is covered in the revised edition of his book *The Art Teacher's Guide to Water-based Screen Printing.*

CASE STUDY: Matt Smith – an eclectic approach

Matt Smith is a good example of what one might call a 'postmodern' approach to the use of printed imagery on ceramics by an artist. Matt is inspired by his time working in the Victoria & Albert Museum and the Science Museum. In the stores of these great institutions, objects are placed together by medium and size, which can lead to interesting and sometimes surprising juxtapositions – like modern-day cabinets of curiosities. In many ways, his work draws on this idea of collecting objects to tell stories with, and he has built up a large collection of moulds and decals, which are brought together to convey narratives, sometimes openly but often ambiguously.

He says of the work, 'Now and again it is refreshing to stop trying to impose order and let random chance take over – letting ideas, experiences and daydreams collide. The objects I make are mementoes of these random coincidences, pinned down like butterflies or photographs and preserving their memories – giving permanence to fleeting, chance events.'

Saving Judy, Matt Smith, 2008. Ht: 34cm (13½in), white earthenware with cast additions, underglaze colours, custom and vintage transfers. *Photo: Gill Orsman.*

CASE STUDY: Ken Eastman and Royal Crown Derby – transforming tradition

The designs produced in industry for ceramics are a rich source of inspiration for artists. The collaboration between Royal Crown Derby and Ken Eastman shows how 'traditional' motifs can be transformed through being applied to contemporary forms.

Ken Eastman and Royal Crown Derby, from the 'Four Seasons' series, 2009. Ht: 15cm (6in). *Photo: courtesy of the artist.*

Notes

[1] For more details on the history of early transfer printing see WILLIAMS-WOOD, C., *English Transfer-printed Pottery and Porcelain* (London: Faber and Faber Ltd, 1981), and COPELAND, R. & SPENCER C., 'Developments in printing as used in the decoration of pottery' in *Trans Brit Ceramic Soc.*, Oct 1955, 583–609. 'Japan' is a varnish of exceptional hardness. 'Japanning' is the action of japanning or varnishing with Japan. See WILLIAMS-WOOD (1981, p.40) for an explanation of the mid-18th century craze for sticking engravings onto japanned ware, which he suggests inspired the idea of printing straight onto the japanned surface.

[2] See HALFPENNY, P. (ed.) Penny plain, *Twopence coloured – Transfer printing on English ceramics 1750–1850* (Stoke-on-Trent City Museum and Art Gallery, 1994).

[3] For more details of history and methods see DRAKARD, D., *Printed English Pottery – History and Humour in the Reign of George III 1760–1820* (London: Jonathan Horne Publications, 1992).

[4] See Copeland, ibid.

[5] See Copeland, ibid.; see BRADSHAW, A., 'Mechanical decoration of ceramics – the way ahead' in *Ceramics Industries Journal*, October 1982, 12–14, & 24.

[6] See GATER, R., 'The development of offset decorating processes in the tableware industry' in *Ceramics Industries Journal*, April 1984, 19.

[7] COLE, D., 'Transfers for automatic application' in *British Ceramic Transactions and Journal*, 89, July/August 1990, 149–152.

[8] 'Innovations in decorating equipment' in *American Ceramic Society Bulletin*, 68, January 1989, 70–71.

[9] BECK, R.C., OWENS, E.A. & STEINMAN, J.H. (1918). *Patent No 1, 254, 764*. Method of delineating or reproducing pictures and designs.

[10] SIMON, S. (1907). *Patent No 756*. Improvements in or relating to stencils.

[11] See HOSKINS, S., 'Screen meshes' in *Printmaking Today*, 4, 1995, 4, 31

[12] WILLIAMS, R, & D., 'The Early History of the Screenprint' in *Print Quarterly*, III, 4, 1986, 287–321.

[13] FREEMAN, P. (1983) Lithographic and screen printed transfers, *Ceramics Industries Journal*. 1992 December, 24–27.

[14] 'Transfer shortage solved?' in *Pottery Gazette and Glass Trade Review*, February 1949, 142–5.

[15] See SALIM, M.S., 'Overview of UV-curable coatings' in *Radiation Curing Polymers II. The proceedings of the third international symposium. Organised by the N.W. region of the industrial division of the Royal Society of Chemistry*, U.M.I.S.T. Manchester, 1990, 3–21; and WEBSTER, G., 'Radiation curing – Where does it fit in?' in *JOCCA*, 1991 (1), 7.

[16] LOGAN, A., *The development and testing of novel on-glaze water-based transfer-printing processes for decorating ceramics*. Ph.D. thesis, University of the West of England, Bristol, 2001.

SHAKSPEARE'S HOUSE HENLEY S.T STRATFORD ON AVON.

3

How ceramic transfer prints look: methods and aesthetics

Printing is a means of reproducing an image on the ceramic surface. But what are the qualities of that image? This chapter links methods and the resulting aesthetic characteristics.

Many artists using transfers will see the process as a straightforward method to translate an image – perhaps a photograph or drawing. However, there are a number of subtleties of transfer printing, that some might explore and develop. This chapter is by no means exhaustive, but I hope it might also be regarded as a source of inspiration, and with this in mind I have used the images to show conventions, quirks and practical aspects that might be utilised, referenced or subverted. The chapter ends with a case study on Charlotte Hodes, who successfully combines approaches in her work. For more detailed accounts you can refer to the books and articles listed in the endnotes.

Aesthetic and practical attributes of transfer-printing from engraved metal plates

Engraving and etching dictated that the image could only be composed of lines and/or dots.[1] So completely solid areas of colour were not possible because, when an engraved or line-etched plate is wiped, any ink sitting in any areas that are widely engraved or bitten will be wiped away and will consequently not print. Areas of tone or near solids can be achieved by stipple engraving or cross-hatched lines drawn close together. During firing ceramic pigment may diffuse slightly to give the appearance of a near solid.

There are two main methods to help distinguish between ceramic prints produced by bat and those produced by tissue.[2] In the former, the flexible glue bat could stretch and conform to the shape of the ware without tearing or producing creases in the image, which are sometimes associated with tissue printing. However, this sometimes resulted in a conspicuous distortion of the image, often more prominent on curved hollowware.

The second indicator is the presence of small unprinted areas sometimes with a central dot. These blemishes are caused by dirt or air trapped between ware and bat, thus preventing full transferral of the image. In tissue printing this does not occur, as trapped air passes through wet tissue when the transfer is applied.

Transfer prints produced by the tissue method have different aesthetic qualities to those of bat printing. On application the tissue can crease, causing a blemish in the image. Also, tissue prints were often used to decorate a large portion or even the

Multi-plate printed pot lid, 19th-century. Dia: 10cm (4in). *Collection of the author. Photo: David Williams.*

ABOVE: Herculaneum plate depicting Nelson. Stipple engraving made by punching the surface of the metal plate with a pointed tool. *The Potteries Museum & Art Gallery, Stoke-on-Trent.*

LEFT, TOP: The distortion of bat printing – 'The storming of the Bastille', 18th-century bat-printed jug. The distortion of bat printing can be seen on the jug's printed border, which has become misshapen. *Photo: © V&A Images/Victoria and Albert Museum, London.*

LEFT: 'The storming of the Bastille' (detail). 18th-century bat-printed jug. Note the white unprinted areas in the castle wall. *Photo: © V&A Images/Victoria and Albert Museum, London.*

RIGHT: All-over tissue-printed decoration on a large jug. Lead-glazed earthenware with transfer-printed decoration in blue & white, made by Bourne, Baker & Bourne for the retailers Bailey & Neale of London. English (Fenton, Staffordshire), 1830. *Photo: © V&A Images/ Victoria and Albert Museum, London.*

whole surface of the ceramic article. This is particularly common on blue and white pieces. Bat prints tended to be self-contained on the surface of the ware in the style of a vignette due to the difficulty of application with the unwieldy gelatin bat.

When a tissue print covers a large portion of the surface of the ware pieces of tissue have to be laid down separately to make up the image. This is necessary because a single printed sheet would not be flexible enough to cover the ware without severe

creasing. The joins between the separate pieces of tissue can usually be clearly seen on the fired article. On some pieces, parts of the decoration can be seen to have been cut away and used to decorate other parts of the ware.

Another aesthetic characteristic of early transfer printing relates to the positioning of the print on the ware. As the same printed image was often used to decorate different size and shaped ware, it did not always fit correctly. For example, on the mug shown above the print is too big and so goes over the rim. On other pieces the transfer is applied at an angle to facilitate the fit. This attribute of some pieces does not relate to an inherent difficulty of the process but to a lack of care when considering the combination of form and decoration. In my view this adds charm to pieces and is an approach that the contemporary artist might consider for certain works.

'Matrimony and Courtship' mug, 18th-century. The faces have two different expressions standing upright and upside down. Note how the print does not fit and goes over the edge of the mug. *Photo © V&A Images/Victoria and Albert Museum, London.*

BELOW: Cut away tissue prints on a blue and white bowl. Collection of the author. *Photo: David Williams.*

The use of colour in early transfer printing, *c.1760–c.1860*

The use of colour presents another aesthetic characteristic of early printed pottery. Very early transfer prints tended to be black or brown. Later, Chinese ware inspired the use of perhaps the most famous colour, cobalt oxide blue, for example in the case of the Willow Pattern.[3] Although the actual print applied to pottery was almost always monochrome, transfer-printed wares were routinely hand-coloured with enamels. In the 19th century, transfer printing from copper plates started to be used to provide an outline for the hand-painters.

From the 19th century there are examples of transfer prints from single engraved plates printed in several colours.[4] At the time the range of colours that could withstand the glaze firing was limited – blue, black and green being the most common.[5] Because of this a full-colour reproduction was not possible; although blue ink could be used for the sky and green for the grass, etc.

Multicolour single-plate printing was soon superseded by the development of multi-plate printing. This process involved the separation of the image to be printed into individual colours – usually yellow, blue, red and black. A separate plate would then be engraved for each colour. The images on the plates would be applied separately to the ware by tissue, to build up the full-colour design.[6]

The multi-plate process can also be seen as a precursor of four-colour CMYK (Cyan, Magenta, Yellow, Keyline) printing, which is one of the standard techniques used to produce screenprinted transfers in industry today.

'Italian Verandah' dish, Davenport (detail, see full image on p.18), *c.1835–40*. Multicolour single-plate print. *The Potteries Museum & Art Gallery, Stoke-on-Trent.*

Detail of multi-plate printed pot lid, 19th century. Note the rich surface that can be created from just four colours – yellow, blue, red and black. Also note the range of colours that can be created by overlapping colours: for example, the orange under the window on the left where red and yellow is overlapped, or green of the lady's dress on the right where blue and yellow overlap. Similar effects can be achieved by overprinting colours in screenprinting. (See also full image on p.28.) *Collection of the author. Photo: David Williams.*

RIGHT: Progression of printing a multi-plate design on a pot lid. 'The Village Wedding', F.R. Pratt & Co., after 1883. Each colour is printed from a different engraved plate using potter's tissue. *The Potteries Museum & Art Gallery, Stoke-on-Trent.*

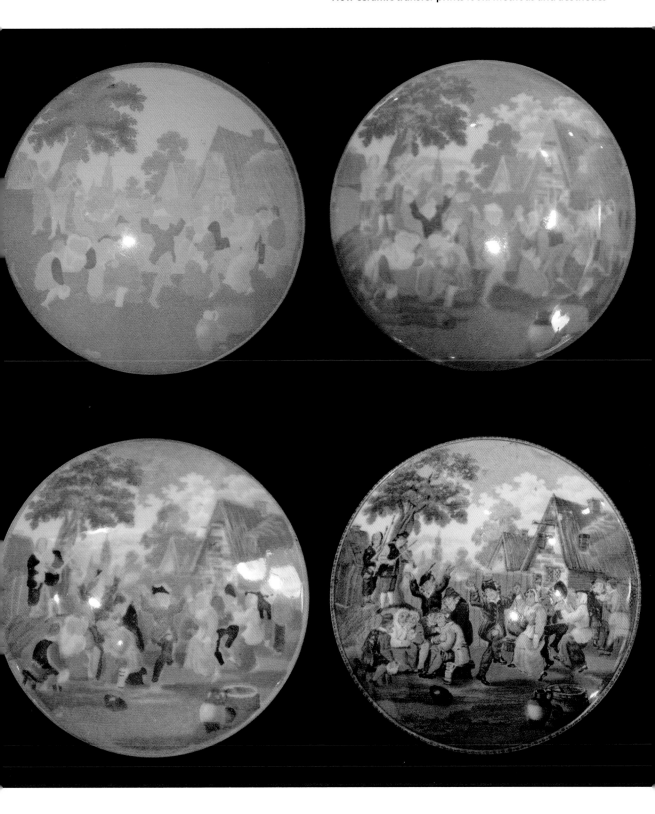

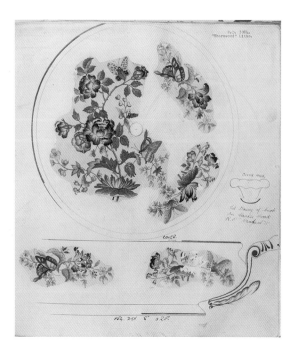

Aesthetic and practical attributes of lithographic ceramic transfer printing

Lithography is not dealt with in detail in this book because the subtle effects that it is known for can now be achieved with screenprinting. However, it might be useful to consider examples as possible inspiration. Lithography is characterised by its ability to reproduce soft shaded effects due to the thin layer of colour it deposits on the paper.[7] It is also less complex than tissue or bat printing, in terms of application, as all the separate colours of an image are printed onto the transfer paper together and then applied to the ware.

In lithography almost any kind of drawn mark can be made as long as the drawing medium is greasy, because the process works on the basis that oil and water do not mix. During printing the surface of the lithography plate is wetted so that the oil based ink only adheres to the drawn areas. These drawn marks may include lines, solids and stipples. These can be seen in work created by Eric Ravilious for Wedgwood in the 1930s and '40s. Ravilious (1903–42) was a successful illustrator, watercolour painter and designer, who produced several designs for Wedgwood.[8]

Ravilious's 'Barlaston' mug of 1940, designed to commemorate the move of the Wedgwood factory from Etruria to Barlaston, was decorated with a four-colour lithograph. The range of marks that can be seen in this design exemplify the greater diversity of image-making approach which is possible with litho compared to engraving. In this piece the image is composed of line, areas of solid colour, wash effects and stipples. The ability to print in several colours negated the need to embellish the image with hand-painted colour.

ABOVE LEFT: Charnwood lithograph pattern-book entry, 1947. This design appears to emulate engraved and hand-coloured patterns, in that the image is composed of a dark keyline with solid and stippled colours which resemble hand-painting. *Image courtesy of the Wedgwood Museum Trust, Barlaston, Staffordshire.*

ABOVE RIGHT: 'Barlaston' mug, designed by Eric Ravilious, 1940. Litho-printed. Note how the overlap of the yellow and grey colours in the flames creates a new shade. This can be achieved with screenprinting today. *Image courtesy of the Wedgwood Museum Trust, Barlaston, Staffordshire.*

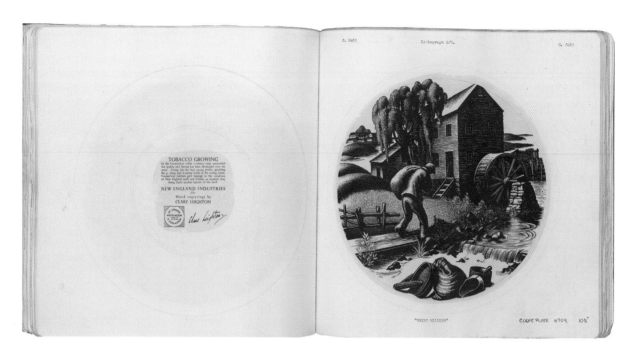

Grist Milling, plate pattern book entry from 'New England Industries', designed by Clare Leighton, 1952. Litho-printed. A good example of how a composition might be developed 'in the round' on a plate. Also note the backstamp – worth considering when you develop your own designs. *Image courtesy of the Wedgwood Museum Trust, Barlaston, Staffordshire.*

The development of photolithography allowed for the photographic reproduction of imagery and consequently extended the aesthetic range of imagery that could be transfer-printed. An early example produced by Wedgwood was Clare Leighton's 'New England Industries' limited-edition plates. Clare Leighton (1898–1989), a notable wood engraver and illustrator, was commissioned in 1952 to design a series of twelve plates for Wedgwood, depicting New England industries.[9]

Leighton's woodcut designs were reproduced underglaze as lithographic transfer prints in charcoal sepia and applied to coupe, meaning rimless, plates in Wedgwood's Queen's Ware. The advantage of using photolithography was that Leighton's woodcuts could be faithfully reproduced. Before the advent of photographic reproduction techniques, Leighton's designs could not have been reproduced so faithfully on ceramic. Again, screenprinting today offers this potential to reproduce imagery from another medium for application to ceramic. For example, a drawing or photograph might be scanned and then screenprinted.

The diversity of screenprinted ceramic transfers

The study of four calendar plates (pp. 38–9) produced by Wedgwood between 1971 and 2004 show the development of transfer printing during the period.

The thicker deposit of ink achieved by screenprinting is several times greater than that possible with lithography. This results in a more intense fired colour and is a key advantage over lithography.[10] In lithography, in order to achieve the strongest possible colour strength the amount of flux in the ink has to be reduced so that more enamel

Wedgwood calendar plate, 1971. Litho transfer-printed. The 1971 calendar plate is produced by lithography. The image is composed of areas of line, solid and also very fine halftone, almost imperceptible to the naked eye. Note how the colour is printed over the black, giving the impression of being hand-painted. In my view this level of subtlety offers great potential to studio artists today using screenprint. *Collection of the author.* *Photo: David Williams.*

Wedgwood calendar plate, 'Tonatiuh', 1977. Screenprinted transfer. Six years later, this design is composed of flat areas of bright solid colour and line work. There is no halftone in this image. Note how in places the flat areas of colour overlap the line work and change its appearance – for example, in the blue and the pink areas. Again, a level of subtlety that could be exploited today. *Collection of the author.* *Photo: David Williams.*

Wedgwood calendar plate, 1984. Screenprinted transfer. This has much less conspicuous areas of solid colour than the 'Tonatiuh' plate. This image also features the use of halftone areas, though these are still crude compared to litho in that the dot structure of the halftone can be clearly seen with the naked eye. Note the range of types of mark. *Collection of the author. Photo: David Williams.*

Wedgwood calendar plate, 2004. Screenprinted transfer. By 2004 the halftone work is quite fine and much less visible than in the 1984 example. Note how the halftone dots give the appearance that this image is a facsimile of an artwork in another medium – perhaps a watercolour. To my mind this image is less interesting than those that exploit the qualities of screenprinting in its own right rather than its ability to reproduce artwork. *Collection of the author. Photo: David Williams.*

'Yellow Daisy' cup designed by Suzie Cooper. Screenprinted transfers. *Photo courtesy of the Wedgwood Museum Trust, Barlaston, Staffordshire.*

can be added. As it is the flux which creates the gloss of the fired transfer, this can result in a sacrificing of gloss in favour of colour strength.[11]

Despite the advantages, early screenprints were limited to bold, simple designs. In contrast lithography was better suited to fine, detailed work in paler colours. However, due to advances in screenprint technology in the 1980s,[12] the range of possible imagery has increased. The generic types of imagery which screenprinting can now reproduce are: solid areas of flat colour; basic line and halftone photographic work; full-colour photographic work (CMYK); hand-drawn imagery; and computer-generated artwork.

The pattern 'Yellow Daisy' combines the use of solid areas of colour with 'halftone' areas which emulate hand-painting. Note the blue and green 'brush strokes' which appear hand-painted but are actually halftone reproduction of hand-painting. The precise straight edge on the green 'brush stroke' on the rim of the saucer is too neat for hand-painting. This relationship between reproduction and hand craft in printmaking strikes me as an interesting area to consider when making images.

Digital transfers

A new generation of digital transfer offers the potential for the reproduction of very fine photographic imagery (see Chapter 9). The combination of a range of methods, such as screenprinting, digital printing and other ceramic decoration techniques, offers the potential for a very rich ceramic surface.

CASE STUDY: Charlotte Hodes – Successfully combining digital and screenprinted transfers

Charlotte Hodes's current work takes the form of large-scale intricate paper-cuts and work onto ceramic. This is informed by what might be called a 'collage methodology': both hand cut-and-paste in the studio as well as digital cut-and-paste on the computer. Her

Pink Shade, Charlotte Hodes, 2007. Earthenware with coloured slips, hand-cut digital transfers of fabric across the top and bottom, screenprinted transfers on the white area. 38 x 36cm (15 x 14¼in). *Photo: Peter Abrahams, courtesy of Marlborough Fine Art.*

subject is the female figure, as a silhouetted motif intertwined with historical depictions of women, pattern, ornamentation and contemporary motifs such as computer icons and kitchen pots. Winner of the Jerwood Drawing Prize in 2006, Hodes is a Senior Research Fellow at the London College of Fashion, part of the University of the Arts, London.

In my view, the work of Charlotte Hodes offers a masterclass in the combination of screenprinted and digital transfers. The collage methodology that Hodes uses in much of her work extends to the way in which she applies transfers onto ceramic pieces. She makes copious drawings and takes many digital photographs which she scans into the computer, reworking them using basic computer functions such as 'cut', 'paste', 'scale', 'filter', etc. These are then produced as digital or screenprinted transfers. The transfers provide her with a huge archive of visual imagery, which she can freely cut and paste, working directly and intuitively, onto the ceramic surface.

The digital transfers enable her to create 'one-off' sheets of motifs or patterns at a reasonable cost, whereas screenprinted transfers allow numerous repeated sheets of the same imagery, which if the screen is cleaned, can be printed again in another single colour. It is only economical to use screenprinting if multiple copies are required, especially when they are not printed by the artist/ceramicist, but, as in her case, are outsourced to a screenprinting studio.

Hodes also exploits the different qualities of the digital and screenprinted transfers. The digital transfers have a thinner deposit of ink thereby producing a weaker, paler image; whilst a screenprinted transfer has a denser deposit of ink, so that a single colour transfer on the ceramic is richer and more 'physical', in that it creates a 'skin' of colour with a slight relief. These contrasting characteristics are used not only to make the imagery pictorially less flat but also to provide a dynamic contrast across the decorated ceramic surface.

In addition, the digital transfers are excellent for recreating continuous tone and photographic images, such as the pink drapery used in *Pink Shade* (see p.41). In this way, Hodes avoids the equivalent, more laborious and costly, four-colour (CMYK) process of screenprinting. *A Dinner Service* utilises the dark black and grey photographic pattern (sourced from a photo of a floral detail in a 17th-century Dutch painting by Jan Van Huysum, which was scanned into the computer and reworked). This contrasts with the hand-cut silhouettes made from sheets of single pastel-colour screenprinted transfers.

A Dinner Service, with 14 place settings, Charlotte Hodes, 2008. Hand-cut digital and screenprinted enamel transfers onto white china. *Photo: Peter Abrahams. Private collection.*

Notes

[1] See WILLIAMS-WOOD, C., *English Transfer-printed Pottery and Porcelain* (London: Faber and Faber Ltd, 1981).

[2] See DRAKARD, D., *Printed English Pottery: History and Humour in the Reign of George III 1760–1820* (London: Jonathan Horne Publications, 1992).

[3] For more detail see COPELAND, R. & SPENCER, C., 'Developments in printing as used in the decoration of pottery' in *Trans Brit Ceramic Soc.*, October 1955, 583–609.

[4] See TILLEY, F., 'The so-called polychrome printing' in *The Antique Collector*, September/October 1946, 183–6.

[5] See HALFPENNY, P. (ed.), *Penny plain, Twopence coloured – Transfer printing on English ceramics 1750–1850* (Stoke-on-Trent City Museum and Art Gallery, 1994).

[6] For an artist's perspective see MASON, P., 'Pot lids to Paolozzi' in Scott, P. & Bennet, T., *Hot off the Press: Ceramics and Print* (London: Bellow Publishing Company Ltd, 1996), pp.15–18.

[7] See FREEMAN, P., 'Lithographic and screenprinted transfers' in *Ceramic Industries Journal*, 92, December 1983, 24–7.

[8] GATER, R., 'The development of offset decorating processes in the tableware industry' in *Ceramic Industries Journal*, April 1984, 19.

[9] From a contemporary publicity leaflet: WEDGWOOD, JOSIAH & SON INC. (no date). 'Wedgwood reproduces New England's basic industries, harvests of land and sea from woodcuts by Clare Leighton.'

[10] For accounts of the transition from lithography to screenprinting see FREEMAN, P. 'Lithographic and screenprinted transfers' in *Ceramic Industries Journal*, 92, December 1983, 24–7; and 'Screenprinting – a decade of growth' in *Ceramic Industries Journal*, 92, April 1983, 14–15.

[11] See FREEMAN (1983), ibid.

[12] See STEPHENS, J., *Screen Process Printing* (2nd edn) (London: Blueprint – an imprint of Chapman & Hall – 1996).

4

Materials for ceramic transfer printing

Many of the processes described in this book are very similar to methods of printing onto paper. However, in order for your prints to be fired onto ceramics there are important differences in the printing mediums and colourants used.

Inks for printing onto ceramics

Ink consists of a colourant, often in powdered form, and a medium. The medium, which can be solvent-based or water-based, acts as a carrier for the colour during printing. The methods described in this book involve the 'fixing' of this colourant onto the surface of ceramics, enamels and glass with heat to form a permanent or semi-durable bond. This heat is usually achieved by placing the transfer-printed piece into a kiln and firing it.

The need for heat to fix the print requires special types of colourant. Standard inks for printing onto paper would burn away. Therefore, vitrifiable colourant must be used. This means that while printing medium burns away, colourant remains and is changed into a glass-like material through the action of heat.

For printing permanent images onto ceramics, glass and enamel on metal, this colourant will most likely be enamel. Enamels are comparatively low-firing glass frits containing metallic oxides as colouring agent, and available in either powder or paste form. The methods given in this book mostly use powdered enamels.

Enamels are also available in either opaque or transparent form, which refers to their appearance after firing; it is also possible, however, to add a transparent flux to opaque enamels to increase transparency. They also come in different firing temperatures. This area can cause some confusion because the lower-firing enamels tend to be used for glass (approx. 550–650°C/1022–1202°F), and the higher-firing for ceramic (approx. 750–850°C/1382–1562°F). This is because, unlike ceramic, glass objects will melt at the higher temperature. Therefore, I would tend to describe the lower-firing range as 'glass enamels' and the higher-firing range as 'ceramic enamels'. This is important to bear in mind when choosing and buying enamels. It would be advisable to check that the firing temperature is suitable for your aims. Underglaze and in-glaze ceramic colours can also be used; these have a firing temperature of around 1280°C (2336°F).

Willow Pattern, Rob Kesseler, 2007. 250 x 12cm (98½ x 4¾in), bone china with printed decals of magnified leaf structures. *Photo: courtesy of the artist.*

Printing mediums

The printing medium that you choose will largely depend on the method of printing, i.e. screenprinting, etching, etc. In screenprinting perhaps one of the main issues will be whether to use a solvent-based or a water-based (more accurately, water-miscible) medium. In basic terms, a solvent-based transfer is perhaps easier to use once made, and will give slightly more consistent results than a water-based transfer. However, the water-based transfer will be easier to make in the first place. For printing from a metal etching plate or flexography plate via tissue you can use the standard copperplate oil used to make etching ink or litho/relief printing ink (see Chapter 10). The suppliers of enamels and mediums will advise on the products suitable to your requirements.

Solvent-based versus water-based for screenprinted transfers

This table gives a general overview of the strengths and weaknesses of solvent-based and water-based transfer printing. In industry solvent-based systems are used. However, for the studio-based artist, water-based systems have the advantage of not giving off so many fumes and being easier to clean up. With care, the artist can also work with and minimise the disadvantages. I have worked extensively with both solvent-based and water-based transfers, but will focus on water-based in this book, largely because as far as I am aware, in-depth details on the process have not been published widely before, and also because, in the UK at least, health and safety legislation is making it harder to use solvent-based transfers in art schools.

TYPE OF TRANSFER	ADVANTAGES	DISADVANTAGES
Solvent-based	The ink takes longer to dry on the screen and so is perhaps easier to print. The finished transfer is more flexible and so is easier to apply to curved objects.	The medium and covercoat require extraction to remove fumes. The fumes can cause headaches. Both ink and covercoat will need to dry overnight, in most cases extending production time to about two days. Requires the printing of covercoat, which can be tricky and extends production time. Requires the use of solvent to clean screens and equipment.
Water-based	The ink dries quickly and the transfer paper is already covercoated, making production time quicker (about a day). Screens and equipment can be cleaned with water.	The ink dries quickly in the screen, meaning that the screen might have to be cleaned up several times during long print runs. The finished transfer is less flexible and so less suitable for compound curves. Cracks can occur in the print where it folds.

Summary of colourants and mediums

This table gives a very general guide to the sorts of printing materials that you might require for different types of ceramic transfer printing. More detail is given in the relevant chapters and in the Directory at the back of the book.

On-glaze, in-glaze or underglaze?

When discussing ceramic decoration in general, the convention is to talk in terms of where the decoration is placed in relation to the glaze. The same is true of transfer prints. This is important because the position of the decoration dictates the colourants that can be used.

Underglaze decoration is applied to the 'green' (unfired) or 'biscuit' ware (this is fired but not glazed ceramic). As the glaze (usually clear) is applied over the decoration, and then fired, you will have to use colourants that can withstand the high temperatures of a glaze firing. These will be underglaze colours or oxides such as cobalt.

On-glaze decoration is applied to the surface of glazed and fired ceramics. These are then fired again at a much lower temperature than the glaze was fired at, but high enough to soften the glaze and the on-glaze colour so that they bond together. On-glaze decoration, usually coloured enamels or lustres, can sometimes be seen to 'sit' on the surface of the glaze.

TYPE OF TRANSFER PRINTING	SUITABLE COLOURANT	SUITABLE PRINTING MEDIUM	TYPE OF TRANSFER PAPER	WHERE DO YOU GET THEM?
Water-based screenprinted transfers	• On-glaze enamel • Underglaze • In-glaze • Oxides	• TW Flat Clear Base • Daler Rowney Glaze Medium Gloss and Acrylic Retarder	U-WET	Colours from ceramic and glass suppliers. TW Flat Clear Base and U-WET from John Purcell Paper Ltd. Daler Rowney products from artshops and www.daler-rowney.com
Solvent-based screenprinted transfers	• On-glaze enamel • Underglaze • In-glaze • Oxides	Solvent-based screenprinting medium for transfers	• U-WET • Gummed transfer paper in combination with covercoat	• U-WET as above. • Ceramic and glass suppliers.
Printing from an etching or photopolymer plate onto ceramic	• Underglaze colours • Oxides (cobalt oxide works well)	• Medium copperplate oil • Litho/relief printing medium	Potter's tissue to transfer the image	Copperplate oils and litho medium are available from printmaking suppliers; potter's tissue is difficult to get at present.
Digital transfer prints	Enamel supplied as part of transfer	Medium supplied as part of transfer	Transfer paper supplied	Digital transfer printers.

In-glaze decoration is applied to the fired surface of the glaze, but underglaze or oxide colours are used. The piece is then fired again at a temperature close to the original glaze firing, which makes the decoration 'sink' into the glaze.

Underglaze and in-glaze offer a very robust decoration and are less likely to wear away, but the range of colours tends to be more muted than with on-glaze enamels, which are available in bright colours. On-glaze enamels are also very tough, though lustres can be prone to wear. In my view, on-glaze transfers are probably those best suited to general use. Underglaze decoration may be preferred for plates where wear from utensils could be an issue.

Transfer papers

There are two main types of transfer paper used for screenprinting and covered in this book: gummed transfer or decal paper for solvent-based printing (often called 'traditional transfer paper' in the US), and 'U-WET' transfer paper, which can be used for both solvent and water-based transfer printing. Both are described in Chapter 1. There is a another type of paper used for transfers that are not fired on to a high temperature (Lazertran), which is not covered but see www.lazertran.com for more details. Also digital transfer printers use a specialist paper but this would be supplied when you order your digital transfers.

Screens

Screenprint screens are relatively cheap to buy. Therefore, it is not necessary actually to make your own frame and stretch the mesh over it. However, if you are especially keen to do this, Paul Andrew Wandless gives instructions on making screens in his excellent book *Image Transfer on Clay* (Lark Books, 2006).

An important aspect to consider when choosing a screen is the mesh size. In most cases the mesh size is denoted by a number and a letter – for example, 120*t*. The number refers to the number of mesh threads per centimetre while the letter relates to the diameter of the mesh thread. The letter *t* denotes a standard diameter, and *HD* means heavy duty. For water-based inks you would usually use a 120*t* for hand drawn (autographic) stencils and a 150*t* for photographic ones. For solvent-based inks a 90*t* is more common for hand drawn stencils and 120*t* for photographic ones. A 120*t* should be suitable for most uses.

Types of ceramic

You can apply transfers to most kinds of ceramic and glaze. One thing to remember is that the surface of the fired transfer is likely to look similar to the surface of the glaze to which it is applied. So, if you apply the transfer to a textured matt glaze, then the final image is likely to take on the same kind of texture. Transfers are perhaps most commonly used on 'whiteware', i.e. white-glazed bone china or porcelain. Many artists and designers buy blank whiteware that they then decorate with transfers. If you do this it is best to go for bone china or porcelain, as earthenware can suffer from 'spit-out' when the transfer is fired on. This is an unsightly, blistered effect caused by moisture leaving the clay body.

Health and Safety

When using any materials or processes in art and design you must take note of Health and Safety guidelines. Control of Substances Hazardous to Health regulations and risk assessments are now a major consideration in any studio. The area of ceramic transfer printing presents some potential hazards for the unsuspecting. For example, many of the printing mediums used may require the wearing of a mask to prevent inhalation of the fumes. This is especially the case with solvent-based products, where you should also ensure adequate ventilation. Gloves should also be worn where necessary to prevent contact with the skin. You should also wear a mask or respirator when handling enamels, mould-making materials and any other dust-like materials. You should follow the guidelines given in the studio that you are working in. If setting up your own studio you will have the legal responsibility not only for your personal safety but also that of workers and visitors.

Avoid eating or drinking in the studio and ensure that materials are cleaned up after use. Specialist guidance should be sought before you use any material that you are not familiar with. Most products will have Health and Safety guidance notes (or MSDS Reports in the USA), which must be adhered to. In most cases common sense is perhaps the best safeguard.

At Royal Crown Derby enamel inks are mixed with a triple roll mill to a homogenous consistency.
Photo: Kevin Petrie.

5

Screenprinted waterslide transfers: the basics

The prime method of ceramic transfer printing is the waterslide transfer or decal. This chapter describes how to make both water-based and solvent-based versions.

An introduction to screenprinting in general

The original screenprint process involved the use of a piece of silk, which accounts for the term 'silkscreen printing'. Because today the mesh is usually made of manmade fibre the terms 'screenprinting' or 'screen-process printing' are more accurate. Other terms have been used, including 'serigraphy' and 'mitography'. The term 'serigraphy' was coined in order to distinguish artists' screenprints from those produced commercially. 'Mitography' is a term devised by Albert Kossloff in 1942, derived from Greek words meaning 'threads' or 'fibres' and to 'write' or 'print'.

For the studio-based artist an important consideration when making screenprinted transfers will be what kind of imagery you want to make and perhaps more importantly what access to screenprinting equipment you have. Most studio-based artists will not have a fully equipped print studio themselves, but they might be able to hire equipment from an open-access studio, local college or university, or from other artists. If you do have access to a screenprint studio you should be able to coat your screen with a light-sensitive emulsion and expose this to UV light with a positive of the required artwork to achieve a resilient and accurate stencil on the screen. This approach is dealt with in Chapter 7.

However, many artists who want to make transfers will not have such access, and approaches in this context are covered in Chapter 6. The current chapter shows the process for making a transfer – assuming that an image is already on the screen – using a light-sensitive emulsion, because I think it is important to explain the principles of transfer printing first. You can then refer to the next two chapters in considering the approach that suits you best.

When I teach classes on screenprinting, I often say that it can be difficult to understand each stage of the process until you have seen it or done it in its entirety. This chapter aims to show and explain the process, but of course there is no substitute for seeing or doing at first hand. So if you are unfamiliar with screenprinting, I would advise that you try and take a class. Another thing I often say when teaching is that when one first sees the screenprinting process it can seem rather long and complicated. However, I really believe that once the basics are mastered making screenprinted transfers is a very quick and relatively easy method.

It is important to note that screenprinting is an art form in its own right, and

Floored, Robert Dawson, 2008. 45 x 45cm (17¾ x 17¾in), print on ceramic tiles. *Photo: courtesy of the artist.*

the methods in this book can be applied to printing onto paper as well as ceramics. Remember that you would need to use ink that is appropriate for paper rather than ceramic ink. You might also want to seek inspiration for your ceramic screenprinting from books on screenprinting for paper.

A step-by-step guide for making a water-based transfer

Preparing to print

The ideal situation for printing a transfer is a standard vacuum hand bench for screenprinting. This firmly holds the screen in place and has a vacuum underneath the bed where the paper is placed. The vacuum pump sucks air through small holes in the bed and holds the paper in position during printing. This is important when printing several colours as slight movements of the paper can cause the various colours to be misaligned on the final print. This form of printing bench is expensive, but small versions are available.

You can also use a simple setup comprising a wooden screen hinged to a board. As there is no vacuum you might need to tape the paper to the baseboard to stop it moving or sticking to the screen during printing. This is OK for printing small numbers of prints, but longer runs would be rather time-consuming. There is also a danger of water-based ink drying in the screen, so you will have to work quickly.

For single prints or where registration is not vital you can just position the screen over the paper and print. You may have to get someone to hold the screen in place. A few small pieces of card should be taped onto the corners of the screen to lift it above the surface of the paper, about 4mm (¼in), (this is explained more fully further on).

The illustrations in this section show a standard vacuum hand bench and the printing of a water-based transfer (see p.55). I am a great believer in carefully setting up prior to printing, as short cuts are likely to cause problems in the long run. Having prepared your screens with a stencil, you now need to tape up areas of the screen not covered by the emulsion. Brown plastic parcel tape is good for this, and helps when cleaning ink off the screen as the tape stops ink getting trapped in the corners. Cut the tape to the length of the inside of the screen, taking care not to damage the screen with sharp scissors. Pulling the strip of tape taut helps fit it into the corner of the screen. Smooth it down so it is as flat as possible. Avoid using several bits of tape that do not stretch across the whole screen as these can get pulled off by the squeegee.

Attach the screen to the printing bench. Position your transfer paper underneath the screen so that the image will print in the correct place – for example, in the middle of the paper. Do not be tempted to try and print right up to the edge of the paper. I would leave around a 2.5cm (1in) gap between the paper edge and print. It might help if you attach the original positive to the paper in the position that you want the print to go, and then look through the screen, pressing down slightly, and position the image on the positive with the corresponding areas of open mesh. Remember to remove the positive before printing. Once the paper is in the correct position you should mark its location by placing three strips of card with masking tape flush against the edges as shown in the illustration on p.55. This is especially vital when you intend to print more than one colour, as you will need to register the paper each time you change the colour to ensure that it is printed in the correct place.

It is now important to set the 'snap'. This is the distance between the screen mesh and the surface of the paper. The mesh should not be touching the paper, as this will cause a blurred image when you print. In order to lift the screen to create the snap distance you should adjust the setting by turning the knobs on either side of the printing bench.

If you are using a wooden screen hinged to a board or just holding the screen over the paper, tape some pieces of card either to the front corners of the screen (all four corners if holding the screen) or onto the baseboard to correspond to the corners of the screen. The snap should be around 4mm (¼in) for a tight screen, a distance you can feel by gently pressing the surface of mesh onto the paper below. If the screen mesh is slack use a greater snap distance.

Once your screen is on the bench, paper registered and snap set, you should place scrap paper (newsprint is good) around your transfer paper on the bed of the bench to block the holes not covered. This focuses the vacuum on your transfer paper and also helps to keep the bed clean should you drop ink onto it. Leave a gap of around 2.5cm (1in) between the scrap paper and the transfer paper. Tape the scrap paper down, as you don't want it moving around when you are concentrating on the printing.

I always like to print a test onto scrap paper before printing onto the transfer paper. This allows me to check that the image is as I want it and that the ink is thick enough. This can help avoid wasting expensive transfer paper. In screenprinting the first print is usually lighter than the rest, so printing the scrap paper first means that your first print onto transfer paper will be stronger. If you have a very detailed image – like a fine halftone, for example – you could just print directly onto the transfer paper in the hope of getting the sharpest possible image.

Mixing the ink

Once the screen is in position you should prepare your ink. Some inks may be supplied ready-mixed, in which case you can use them straightaway. However, if you have powdered enamels, underglazes, in-glazes or oxides, you will have to mix them with a printing medium.

The manufacturers of specific printing enamels will supply information on recommended mixing ratios for enamel and medium. I use TW Flat Clear Base as a 'water-based' medium, and tend to mix the ink by 'feel' as opposed to measuring quantities. After a little experience you will be able to judge the correct quantities. You should add the enamel powder to the medium and stir it thoroughly. The consistency of the ink should be quite thick, perhaps like yogurt. Remember that if the ink looks thin on the paper after you have printed it, it will still look thin once you have fired it onto ceramic. If this is the case, you might want to clean the screen down, add more enamel to the ink, and start again. This will save you time in the long run.

If you are only printing a small number of transfers, there is no need to mix a large quantity of ink. Enough for a generous line of about 2.5cm (1in) wide corresponding to the length of the squeegee will be adequate. Use a plastic spatula to mix the ink, as this can then be used to clean the screen after printing. Metal implements should not be used on the screen mesh as they can split the screen.

In industry the ink is usually mixed through a triple roll mill to create a homogenous mix. For the small-scale user a glass muller and a sheet of sandblasted glass can be used to mix the ink. A pestle and mortar is another quick option. However, in my view none of the above are strictly necessary unless you need to achieve a very precise flat area of colour where spots of unmixed enamel might cause a distraction.

Experienced printmaker David Fortune uses Daler Rowney products to make water-based transfers using the following recipe and guidelines. For on-glaze printing and enough for about six A4 decals:

1 Start with one heaped teaspoonful of Daler Rowney Glaze Medium Gloss.

2 Add one flat teaspoonful of liquid acrylic retarder and mix this small amount together.

3 Now add three heaped teaspoons of powdered on-glaze enamel. **Caution: The on-glaze enamel is dangerous in powder form. It is much safer when wet. Add the powder in the open air, or use an extraction box, and be very careful. DO NOT breathe in any of the enamel powder.** Mix together thoroughly. You will notice that it is a little more gelatinous than acrylic inks for paper. The final mix by volume is approximately 70:30, enamel to medium. This ratio can vary depending on the colour of enamel you are using.

As with all materials you should follow the manufacturer's health and safety guidelines. **With enamels you should wear a mask to avoid breathing in the powder, and you should also be careful not to leave enamel on work surfaces where others might pick it up on their hands and ingest it.**

Printing

You are now nearly ready to print. If you are using TW Flat Clear Base water-based ink, there is an important last step. This ink dries very quickly due to cross-linking of the polymers in the medium. The application of a mild alkali to the screen helps to slow this down. Mr Muscle, the proprietary kitchen cleaner, works well. Spray the screen lightly on the top and bottom, taking care not to wet your transfer paper. Then wipe dry with a paper towel.

To print, pour a pool of ink along the bottom width of the screen, which is slightly longer than the length of the squeegee. Take the squeegee and push the ink forward to the top of the screen whilst holding the screen off the area to be printed so that the paper is not printed. This is called 'flooding the screen' and is important because it deposits an even coating of ink into the mesh and helps prevent the ink drying in the screen between prints. Having flooded the screen, put it down and place the squeegee

LEFT: The screen on the vacuum bed ready for printing. Note that transfer paper, ink and squeegee should all be to hand before printing begins.

RIGHT: To print, lower the screen, push the squeegee down to make contact with the transfer paper below and pull the ink forward to print. *Photos: Megan Randall.*

blade behind the line of ink at the top of the screen. Hold the squeegee at an angle of about 60 degrees and pull forward. This will push the ink through the mesh and onto the paper below. Lift the screen and flood to prevent the ink from drying in the screen. Quickly remove and check the print. If you are satisfied with the image, and are printing more than one sheet remove the printed piece to somewhere safe, place another in registration and continue, remembering to flood after each print stroke.

When you have finished printing place some scrap paper underneath the screen, gather as much ink as possible and pull it with the squeegee to the front of the screen. You can then lift most of the ink up off the screen and onto the squeegee. Return the ink to the pot where it may keep for a short while if covered. Then clean down the screen. Mr Muscle with water is good for cleaning water-based inks. You might find that the screen looks a little stained towards the edges of the print area; don't worry, as this usually disappears when you wash off the stencil with a stencil remover and a power washer. Don't forget to clean the squeegee as well. It is important to clean your screen and tools quickly, as the ink dries very quickly and might not be removable once dry.

The key differences when making a solvent-based transfer

Printing covercoat

When using solvent-based mediums it is important to use a studio with the appropriate extraction facilities. If you are working in an educational establishment or open-access print studio, the technical staff will advise you. If suitable extraction is not available, it is possible to print basic transfers outside using a hinged screen attached to a baseboard.

The ink should be mixed in a similar manner and to the same consistency as with water-based printing, but you should use a solvent-based medium for transfer printing. After printing, the screen should be cleaned with a solvent. The manufacturer of the medium will recommend the appropriate cleaning agent.

Once the prints are dry you can covercoat them. To do this you must first make another screen stencil that will completely cover your image with about a 5mm (¼in) overlap. You can make the artwork for this by painting black ink or photo opaque onto drafting film, True-Grain or Mark Resist, or by cutting the shape out of card. This is then used to make a screen stencil as described in Chapter 7. You then screenprint the covercoat over the top of your print having carefully registered it first. You might find that more snap is required. If you were not using a vacuum bed it would be advisable to tape your paper down, as covercoat is rather sticky and can stick the paper to the screen. If this happens, the solvent in the covercoat can cause the ink of the print to dissolve a little and mark the covercoat screen, and this will be offset onto the next print, so it might be best to clean down and start again.

If you only have to covercoat a few prints you can roll the covercoat on thinly with a small sponge decorator's roller. However, this is a rather crude method, so for longer runs it is advisable to print the covercoat.

Cleaning off the stencil

When you have finished with a stencil it should be removed with an appropriate de-coating chemical. Specially produced de-coating chemicals are used for this job. The supplier of the emulsion that you use will be able to advise you of the most appropriate de-coating agent.

Place the screen in a washing trough and spray with water on both sides. Spray on the de-coating chemical and leave for two minutes. Gently wash away excess emulsion and chemical and then blast with a high-pressure hose to remove any stubborn areas (wearing an apron, mask, and eye and ear protection). If the stencil is hard to remove you could reapply the de-coating chemical and leave for longer. Once clean, the screen should be thoroughly dried before recoating. It can be dried in a special drying cabinet or placed in front of a fan heater. Be careful not to overheat the screen as it may split the mesh. Screens can also be left at room temperature to dry. If you have used a solvent-based ink you will have to degrease the screen after removing the stencil. Screenprinting suppliers will advise on the appropriate products for this, although vinegar can be used as a basic degreaser.

How to apply a waterslide transfer

The application of water-based and solvent-based transfers is very similar. After printing, it is best to leave transfers to dry overnight, although with care some water-based transfers can be used very soon after printing. Cut out the transfer leaving about 4mm (¼in) around the image.

LEFT: After wetting the back of the transfer, wait a few minutes until the covercoat layer and image separate from the backing paper.

RIGHT: Whilst holding the transfer in position, gently slide the backing paper from underneath.

First wet the transfer by placing it under the tap and wetting the back or by dipping it into a bowl of water. Then leave it for a few minutes. If you are going to apply several transfers, they can all be wetted at once and left on a plate.

Do not leave a transfer floating in a bowl of water as the residue of the gum layer will be washed away and the transfer may not stick to the glazed surface so well. Once wet, a transfer will curl up; when it flattens out again it should be ready to transfer. After a few minutes you should be able to hold the transfer between your thumb and forefinger and gently move the covercoat and image layer away from the backing paper. Do not completely remove the backing paper until you are ready to apply the transfer, as it is very flimsy and will be difficult to handle.

Make sure that the object to be decorated is perfectly clean. Methylated spirits (solvent alcohol) can be good for this, but hot water is just as good. Wet the surface of the object with water and then place the transfer image side up in the approximate position that you want it. Gently slide the backing paper out from underneath the covercoat/image layer. You can then correctly position the transfer.

Wetting the surface of the object prior to applying the transfer allows the transfer to slide around. Heating the glazed piece before you apply the transfer helps with it's elasticity. Once it is in the correct position the excess water underneath the transfer should be removed. This can be done by gently smoothing down the transfer with a rubber kidney of the kind used in ceramics. Working from the centre outwards make sure that the transfer is in full contact with the surface of the glaze and that all air bubbles are removed. Any remaining moisture can be removed with a paper towel.

You should avoid creases in the image area as these may show up once fired. Creases left in the clear areas of the covercoat are fine as the clear backing burns away when fired. Leave for an hour or so until dry. The ceramic piece is now ready for firing.

Possible firing schedules

These firing times and temperatures are for bone china/porcelain and may vary for different glazes. If the transfer looks matt when it is fired, it could be underfired, so you could try firing it again at a higher temperature. If it is blurred, shiny and the colour has faded, it could be overfired. Unfortunately, this cannot be remedied.

For black and colour on-glaze:
1st Ramp: 80°C (144°F) per hour to 80°C. 2nd Ramp: 100°C (180°F) per hour to 760°C (1400°F), followed by a soak at 760°C (1400°F) for 1 hour. Orton Cone: 016 to 7.30–8.00 o'clock. Blacks can often be fired much higher – say 850°C (1562°F),

For on-glaze reds:
1st Ramp: 80°C (144°F) per hour for 1 hour. 2nd Ramp: 100°C (180°F) per hour to 740°C (1364°F), followed by a soak at 740°C (1364°F) for 1 hour 20 min. Orton Cone: 017 to 7.30–8.00 o'clock.

For underglaze colours:
Apply the transfer to bisque ware. A coating of shellac can help adhesion of the transfer. Fire the transfer at around 600°C (1112°F) to remove covercoat and medium. Glaze the piece and fire the glaze as you would normally.

For in-glaze:
Print the transfers with oxides or underglaze colours and apply to glazed ware. Re-fire the piece to the glaze temperature or just below.

CASE STUDY: One-day workshops

The speed of transfer production offered by water-based methods means that they could be useful for one-day workshops as well as for the busy individual artist/designer who needs quick results in the studio. I have run transfer-printing workshops for schoolchildren where we have started at 9.30 in the morning and finished at 3.00pm with transfers designed, printed and applied to mugs. They then have to be fired overnight, but the speed of printing is very quick.

I often use a 'self-portrait' theme or a 'self portrait that says more about you than just what you like'. I might start these sessions by asking students to make quick large-scale drawings on paper before making their transfer designs. This can help to build their confidence with drawing.

Rachel Leatherland from The King's Academy, Middlesbrough, showing her 'self-portrait' mug just designed, printed and transferred during a one-day workshop.

BELOW: When designing transfers it is important to remember that one of the advantages of transfer printing over direct printing is that you can decorate all surfaces of ceramic objects. Here 15-year-old Nathan Falcon has cleverly printed a design of the front and back of his head on the outside and inside of a mug. *Photos: Kevin Petrie.*

6

Some 'low-tech' approaches to using screenprinting

Some often overlooked methods can be used to produce screenprinted transfers with a very basic level of equipment.

LEFT: *Portrait plate*, Kevin Petrie, *c.*1992, monoprint transfer on earthenware. Dia: 20cm (8in). This piece was made when I was a BA student. *Photo: courtesy of the artist.*

Even if you do not have access to a fully equipped printing studio you can still make images using screenprinting. You won't be able to make photographic or computer-generated images, but these can be rather banal in any case. Limiting your scope to so-called 'basic' techniques might even result in more creative artwork!

As you will see in the pictures below, many of these methods can be combined together. You could also combine them with other print or ceramic decoration techniques. Please note that all the 'process' photographs for the mugs in this chapter show the use of water-based transfer printing (TW medium and U–WET transfer paper).

Flat sheets of colour

Simply by masking out your screen with parcel tape you can create a stencil to print solid sheets of colour onto transfer paper. You could use this to apply blocks of colour to ceramics or to make collage designs with different colours. Collaged

RIGHT: Printing a sheet of colour. Here the screen is set up on a basic hinged system and has been taped out to produce a sheet of blue printed with water-based ink on U–WET paper. Once dry this colour can be cut up and collaged on to ceramic.

FAR RIGHT: A blend of colours. Here blue and yellow ink were placed on the screen to create a blend. The colours were mixed slightly with the squeegee prior to flooding the screen. The more prints that you make, the more blended the colours will become. *Photos: Megan Randall.*

 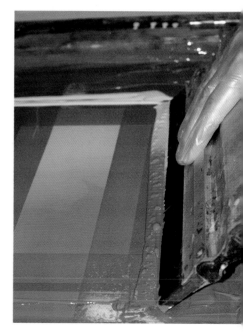

transfers have a tendency to curl at the edges when applied to the ceramic surface. A little gum arabic applied underneath can help stick the transfer down in these areas.

A variation on this is to print a blend of colours. To do this, place two different colours (or more) on the screen and mix together in the middle with the squeegee – or for less of a blend don't mix. Then flood the screen and print as normal. The more prints you do the more the colours will blend together.

CASE STUDY: C. J. O'Neill – adding new meaning to old ceramics

Inspired by memories, ceramics, patterns and graphics, Irish artist and designer C. J. O'Neill works with symbol and pattern to add new meanings to found ceramics. For example, she will take vintage pieces found in charity shops, flea markets or car boot sales, and redecorate them by applying transfers over the old patterns. A key strategy is to use silhouettes as a way to simplify what she sees and convert ideas, thoughts and drawings into surface decoration. To create *Scherzer* she manipulated imagery to design silhouettes of flowers, which were then painstakingly cut by hand from transfer-printed sheets of solid colour and applied to the surface of vintage German porcelain from charity shops.

O'Neill says of her work, 'I hope to reinvigorate existing pieces of ceramics with a new story, allowing them to be cherished for another lifetime. Perhaps in the future they will again be redecorated, added to once more, and become heirlooms of the future.'

Scherzer teacup and saucer from the 'New Heirlooms' collection, C. J. O'Neill, 2005. Hand-cut transfer in red on found gold-printed porcelain. *Photo: Ade Hunter.*

RIGHT: *Profile*, C. J. O'Neill, from the 'Wesley' collection, 2008. Hand-cut transfer in red prior to application on the plate.

BELOW: *Profile*, C. J. O'Neill, from the 'Wesley' collection, 2008. Hand-cut transfer in red on found ceramic. Note how the new red transfer, combined with the existing cadmium in the vintage plate, has reacted, creating a 'burst' effect in some areas. *Photos: Stephen Yates.*

Monoprinting

For a painterly image, monoprinting is an effective technique. You simply paint ink directly onto the screen using brushes or a palette knife (being careful not to damage the mesh). When you have finished the 'painting' you squeegee the ink through the mesh and onto your transfer paper. The ink will blend together somewhat to create a loose, painterly effect.

Remember that if you leave areas of mesh open – i.e. with no ink painted on them – then the ink above this area on the screen will be pushed down and through the mesh with the squeegee. So if you don't want a blurring of the image you must fill all areas of the mesh with ink. If you want areas of your print to remain clear – i.e. with no colour on the print – then simply paint clear printing medium in these areas. You must work quickly, especially if using a water-based medium, as the ink will soon start to dry in the screen.

This method is perhaps best suited to a rapid, spontaneous approach. Some might say, if you only get one print, then why not just paint directly onto the ceramic? Well, of course, you could, but a monoprinted transfer offers a flatter, more graphic aesthetic that is very different in appearance from hand painting.

You might also think about combining monoprinting with other ways of making artwork. Perhaps a monoprinted background area, with a drawing or photographic image printed on top? Also bear in mind that you can use monoprinting techniques through a photographic stencil as a method of getting several colours into an image without making separations on different screens (see p.13). This is most useful if you only need to print a few transfers. For runs of perhaps more than ten, it might be better to make colour separations as described in the next chapter.

Basic stencils

A stencil is essentially something that blocks the mesh of a screen and prevents the ink passing through. You can do this very simply by painting or drawing something onto the mesh that will resist the passage of the ink. For example, if using water-based ink you could paint cooking oil onto the screen. This will block the mesh and because it is greasy will repel the water-based ink, but the oil will of course break down after several prints. The result is a kind of negative image. It will only last for a few prints but could be a quick, easy and effective method for some.

Drawing on the screen with wax crayon is a similar approach but should last for more prints. Try making a rubbing directly onto the screen. A coin works well – but be gentle, as this could also break the screen! Wax can be cleaned out of the screen with turpentine or a similar spirit. Wash the screen well in hot water and detergent to remove the solvent.

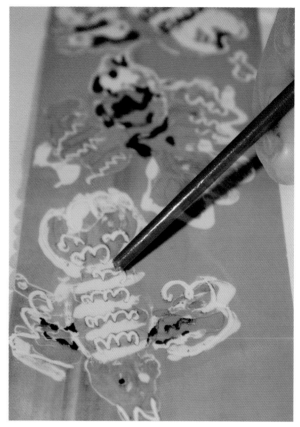

ABOVE: Making a monoprint. Here the ink is made using TW medium. This dries quickly, so it is vital to work fast. If you wanted the fish on a white background you could paint white ink around them or print them using clear medium. Here the background will be a blend of blue and green. Unlike the standard blend method described above it is important not to flood when monoprinting as doing this will disrupt your image.

ABOVE RIGHT: Before the ink dries on your monoprint you can draw back into it (perhaps with the end of a paintbrush). This will leave white lines in the image.

RIGHT: The fired monoprint on a mug.
Photos: Megan Randall.

A wax stencil. Here wax has been drawn onto the screen to block the mesh. Therefore the yellow areas of open screen will print. You can see where a rubbing has been made from a coin (on the left) and from some lace (on the right).

A fired mug with a blend underneath and the wax-resist print from the screen above in black on the top.

A fired 'cooking oil' print mug. First a block of yellow was printed, then the purple was printed on top using the cooking-oil resist method to create an almost textile-like effect.

ABOVE: A hand-cut stencil taped to the back of a screen. The other side of the screen has been taped up with brown parcel tape to ensure that ink only passes through the hand-cut areas.

RIGHT: The printed hand-cut stencil, which has been printed onto one of the blends (blue and yellow) described above. Note that black and purple ink has been used and a 'wavy' blend has been created by moving the squeegee from side to side during flooding. You can see the blend of purple ink (left) and black ink (right). *Photos: Megan Randall.*

Paper stencils

A paper stencil is another easy method of masking areas of the screen mesh to create an image. This method is useful if you need to accurately reproduce a small number of prints – perhaps up to ten. The image is first drawn or traced onto a piece of paper – greaseproof paper or Folex Matt Laser Film works well. You then cut away all the areas that you wish to print with a sharp knife. Then tack the stencil onto the bottom of the screen with some masking tape to hold it in position (*see* above left) and print.

Screen drawing fluid and removable screen block

This two-part method is another simple way to make a stencil on a screen. You simply paint the drawing fluid onto the screen. This is a 'positive' method, unlike the wax resist described above, because where you paint the fluid is what will print. Once the fluid is dry you block out the rest of the screen with removable screen block. Once this is dry the drawing fluid is washed out with cold water to create a stencil. The screen is then printed as usual. The screen block can later be removed with warm water.

ABOVE: Painting an image on the screen with drawing fluid.

ABOVE RIGHT: When the drawing fluid is dry the rest of the screen is coated with removable screen block using a squeegee.

RIGHT: Here the image made with drawing fluid has been printed onto a blend of blue and yellow.

When you come to the end of printing an image, you might like to try turning the paper upside down (or moving it in some other direction) and printing the image again. This overlapping of the image will disrupt the first print if the ink is still wet but may also create interesting effects. A more exaggerated effect can be created by printing the same image many times. Here the rather benign hand-cut stencil of fish (see p.67) becomes a 'frenzy of piranhas' simply by moving the paper through 360 degrees and printing the images several times. *Photos: Megan Randall.*

'Custom run' and 'open stock' transfers

If you do not have access to printing facilities or you need a large number of transfers, perhaps for production ware, you can get a firm to make custom sheets of transfers for you. To do this you would either send to the company an image on paper – maybe a drawing, photograph or lettering – or artwork on disk. They will then make the screens, print the transfers and send you back sheets that you can cut out, apply and fire yourself. You will have to specify what sort of enamel you need the image to be printed in. The more sheets you get printed the cheaper this option will be. For one or two sheets it can be rather expensive.

As well as getting transfers made of your own artworks, you can also buy ready-printed designs from manufacturers. These are usually of a decorative nature, floral designs being especially common. Catalogues are available that you can select from. The designs can then be applied to the objects of your choice. Such open-stock transfers are normally used to decorate production ware, but can also be used for more ironic or subversive purposes. For example, you might consider cutting up and making a collage of open-stock transfers to create surprising juxtapositions. (See Matt Smith's work on p.25.)

CASE STUDY: Robert Winter – improvising in India

In 2008, artist and University of Sunderland ceramics technician Robert Winter was asked by fair-trade company Traidcraft to lead a ceramics workshop at the Sri Sivam Pottery in Pondicherry, South India. The workshop concentrated on design and technical development and the use of on-glaze printed transfers. This proved problematical in practice. The pottery uses traditional wood-burning kilns for low firing, with the result that subsequently the kiln atmosphere was unclean and the temperature inconsistent. Fly ash and other particles would adhere to the transfer surface, giving an unsatisfactory finish. Despite this setback the pottery was determined to add print-decorated pieces to their range, and to this end enlisted the help of small business Inner Reflections, specialists in printed products and packaging.

On first inspection the print shop appeared rather basic. But the finished products on view were faultless, and there was also a computer complete with Photoshop hidden away in one corner. The screens were produced within the hour, but how to get the design onto the pot without transfer paper? Simple – the printer rolled the pot onto the flooded screen. To him it was obvious! The technique is limited to straight-sided vessels but the finished application could not be faulted. A whole new range of products would thus benefit both businesses.

RIGHT: An innovative method for washing a screen in Pondicherry, southern India, when no spray is available. *Photo: Sarah Gee.*

A simple way to transfer an image from a screen to ceramics if no transfer paper is available. First flood the screen with ink, then roll the ceramic over the stencil on the bottom of the screen. Pondicherry, South India.

A bird design printed directly from the screen in Pondicherry, South India. With this method any unwanted areas can be wiped away before firing. *Photos: Sarah Gee.*

CASE STUDY: Charlotte Hodes – combining transfers with other ceramic decoration techniques

Charlotte's work is relevant in this chapter as it shows how simple techniques like using paper stencils to create layers of slip-painted designs in conjunction with transfers and other decoration techniques can create complex and rich ceramic surfaces. 'Vase on a dish' reveals the raised physical surface of the shapes and patterns made by multiple layers of slip-painted-through hand-cut paper stencils, which contrast with the flat patterns made from screenprinted transfers.

Vase on a Dish, Charlotte Hodes, 2007. Dia: 40cm (15¾ in), underglaze and screenprinted enamel transfers onto earthenware, *Photo: Peter Abrahams, courtesy of Marlborough Fine Art*.

CASE STUDY: Paul Andrew Wandless – an alternative to screenprinting, laser toner transfers/decals

US-based artist Paul Andrew Wandless has printed an image from a photocopier onto standard transfer paper (the sort that requires covercoat). The iron in the toner can then be fired onto the ceramic. However, not all photocopiers can accomplish this task, so you'll need to test the machine first. Moreover, the print will need to be covercoated before it can be transferred. See Paul's book, *Image Transfer on Clay*, Lark Books, 2006, for more details and other ceramic and print ideas.

Prodigal King, Paul Andrew Wandless, (back view), 2006. 53.5 x 35.5 x 40.5cm (21 x 14 x 16in), low-fire clay with multi-fired slips, glazes, underglaze chalk and laser toner decal. *Photo: courtesy of the artist.*

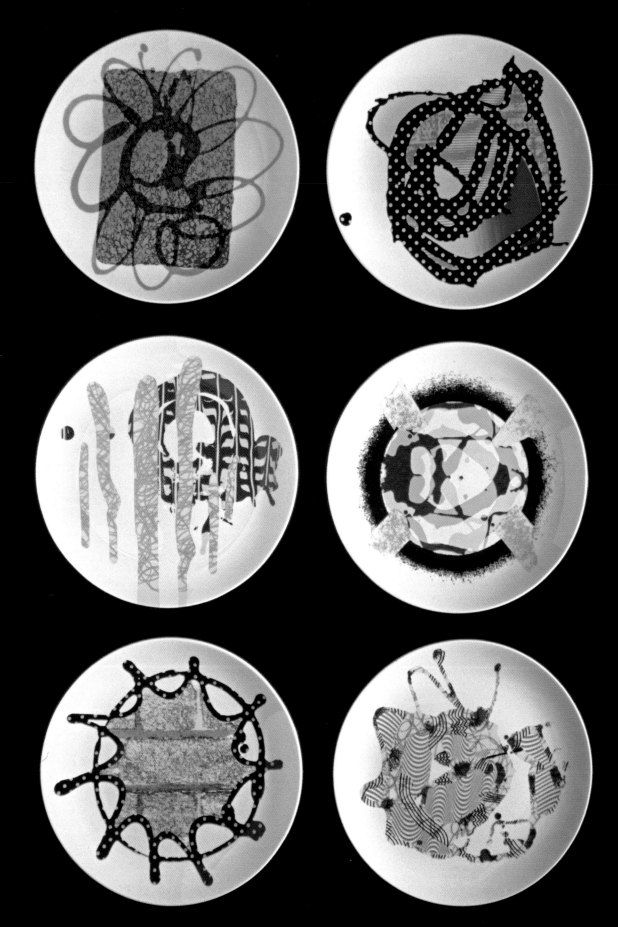

7

Extending the potential of screenprinting: 'photographic' stencils

The ability of transfer printing to reproduce a diverse range of imagery onto ceramic is perhaps one of its key attractions. This chapter describes the prime methods of using light-sensitive emulsion as a stencil.

Photographic stencil production

To make a screenprint using light-sensitive emulsion the mesh of a screen is coated with the emulsion and exposed to ultraviolet (UV) light over a positive of the image required. The UV light hardens the areas of emulsion, which receive light and are blocked by the areas delineated by the positive of the image. The stencil is developed by washing it with water, which removes the coating from the areas which have not been light-hardened, thus leaving the screen mesh open.

This chapter is not just about printing photographic images, although that is an option; rather it considers basic steps relating to the use of light-sensitive emulsions. My personal view is that the photographic image can become rather banal when applied to ceramics, and for that reason I have perhaps focused on hand-drawn effects. The basics of transfer printing were covered in Chapter 5, and this chapter extends the subject with a case study of the printing of an abstract hand-drawn design in several colours. Other case studies offer examples of the approaches of a range of artists.

Developing suitable artwork for screenprinting

You can produce hand-drawn positives by drawing or painting onto semi-transparent films. These films, which can be textured or smooth, can create a broad range of aesthetic effects, ranging from sharp lines to subtle wash effects, when used in conjunction with crayons, paints, inks and other media. The media used for the drawing or painting must be opaque and so able to block out the UV light during exposure. Different positives have to be drawn for each colour. Mark Resist and TrueGrain are both suitable films for drawing on for textural work. Drafting film has a smooth surface and is good for pen work and fine, detailed drawings.

Photographic positives can be divided into two kinds: those which are high in contrast and only recognize black and white as opposed to grey tones, and those which can represent tones through the use of a dot structure. The former is known as a line positive and the latter a halftone positive.

Six plates, Kevin Petrie, *c.*1997. Dia: 25cm (9¾in), screenprinted water based transfers on ceramic. *Photo: courtesy of the artist.*

Notes on preparation of images for print using Photoshop for screenprinting

1. Shoot/scan your photograph or drawing

It is best to input original photos and artworks through scanning. Utilising the scan commands either through Photoshop or your own scanner, scan your image with the resolution set to 300 ppi/dpi (pixels per inch/dots per inch), and then scale no bigger than the size of the paper that you intend to print the positive onto and the size of the transfer paper. Save as a CMYK tiff or jpeg image (8-bit setting). Every scanner is different, so always check for the above settings. Any digital camera will be able to create an image suitable for screenprinting. It is preferable to shoot the images at the highest resolution possible.

2. Open the image in Photoshop

Once you have your image, open it in Photoshop.

3. Resize

Crop your image using the crop tool. In this function you may change image dimensions as well as resolution. For printing, your dimensions should match the size of the final printed object/transfer, with 300 ppi. Remember – it must fit your transfer paper and screen size!

4. Save as ...

Save your new image under a new name, as a CMYK image, if you haven't already done so.

5. Image selection

Choose the part of the image that you wish to use, through the marquee, lasso and magic-wand tools. Remove all unwanted information by selecting the inverse and erasing. You may also use the eraser tool to clean up your image. You may want to do this on a copied layer, in order to preserve your complete image.

6. Colour adjustments for digital transfers

If you want a full-colour photographic image you may decide that digital transfers are an easier option than screenprinting (see Chapter 9). A variety of tools can be used to modify your coloured image, including the curves tool. In general, select the whitest white, the darkest black, and adjust the image as you see fit. Flatten (in the layers window) and save. This image is now suitable for making digital transfers for glass and ceramics (see the Directory at the end of this book).

7. Greyscale images for screenprinting

To make a positive for screenprinting you need to change your colour image to greyscale in Image/Mode menu. To simulate continuous tone, change the mode to Bitmap. Keep the resolution at 300 ppi, then select Halftone Screen. Lpi (lines per

A range of positives. Clockwise from the top: hand-drawn on TrueGrain; a scan on Folex; text on Folex; a linework scan of a cockerel on Folex. *Photo: Megan Randall.*

inch): 10–90, giving a range of dot sizes. Angle: 53 degrees (not 0 or 90). Shape: ellipse (or line).

8. Simple line art

For starters, you can achieve the effects of line art by using two tools:

a) A sketch/stamp tool, which produces a black/white line rendering of your original artwork. But bear in mind that the image smoothness and light/dark balance are variable.

b) Bitmap/50% threshold, which produces a simple black/white image.

9. Incorporating images

Using Illustrator/InDesign, you can incorporate your images into layouts with text and pattern.

10. Printing artwork onto a film

Print your artwork onto a film like Folex Matt Laser Film (FOLAPROOF LASERFILM/F). Artwork can also be photocopied onto film – but do check that the film is suitable for use in a photocopier beforehand so as not to damage the photocopier. This positive is used to make a screen in the same way as a hand-drawn positive (see information following).

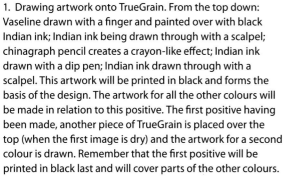

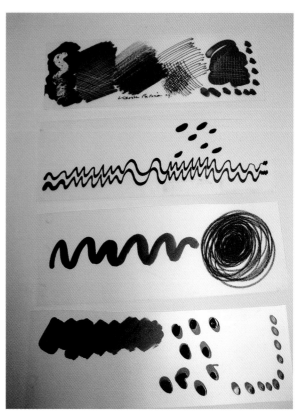

1. Drawing artwork onto TrueGrain. From the top down: Vaseline drawn with a finger and painted over with black Indian ink; Indian ink being drawn through with a scalpel; chinagraph pencil creates a crayon-like effect; Indian ink drawn with a dip pen; Indian ink drawn through with a scalpel. This artwork will be printed in black and forms the basis of the design. The artwork for all the other colours will be made in relation to this positive. The first positive having been made, another piece of TrueGrain is placed over the top (when the first image is dry) and the artwork for a second colour is drawn. Remember that the first positive will be printed in black last and will cover parts of the other colours.

2. The above process is repeated for all the positives. I developed the design as I made each positive, deciding that I would print in the following order from the bottom: yellow, pink, blue and black. Of course you don't have to make the last colour black, and you could also print colours on top rather than underneath – the order of printing will be for you to decide! However, if you have a halftone photographic image it can be useful to print that in black and to make positives for colours to be printed underneath.

CASE STUDY: Printing a 'hand-drawn' abstract transfer for a mug

This case study shows the printing of an abstract design in four colours using hand-drawn artwork and light-sensitive emulsion as a stencil. It is important to remember that although hand-drawn positives are used here, the same principles could be applied to photographic images like the ones shown above. A combination of hand-drawn and photographic images can be very effective.

3. Having made my artwork I now need to coat my screen with light-sensitive emulsion. The four designs can comfortably fit onto two screens, and so two is what will be coated. First pour the light-sensitive emulsion into the coating trough. Offer up the trough to the screen and tip it slightly until the emulsion is in contact with the screen. Then draw the trough upwards to coat the screen. When you reach the top tip the trough back to let the emulsion run back into it, and then draw it away from the mesh. This helps to prevent drops falling down onto the screen. Remember that the emulsion is light-sensitive, so keep the lid on the pot and work quickly. If you have access to a screen-drying cabinet, place the screen in there to dry. If not, put it in a dark room perhaps with a fan heater. Do not leave it out in the light. Return the unused emulsion to the pot. A rubber kidney is useful to help get all the emulsion out of the trough and back into the pot. Remember to wash the trough well after you have finished, as dried emulsion is hard to remove.

4. Once the screen is dry, place your positives on an exposure unit with the screen on top. Position the positives so they are not too close to the frame of the screen, as this will make them hard to print – around a 13cm (5in) gap is good especially at the top and bottom as there is space for the ink. *Photos: Megan Randall.*

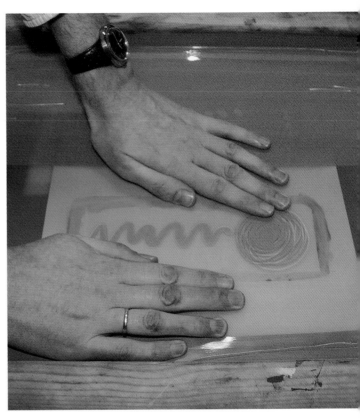

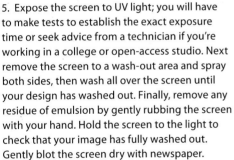

5. Expose the screen to UV light; you will have to make tests to establish the exact exposure time or seek advice from a technician if you're working in a college or open-access studio. Next remove the screen to a wash-out area and spray both sides, then wash all over the screen until your design has washed out. Finally, remove any residue of emulsion by gently rubbing the screen with your hand. Hold the screen to the light to check that your image has fully washed out. Gently blot the screen dry with newspaper.

6. Set up and print the first colour as described in Chapter 5. Then, once the first colour is dry (this will be very soon if you are using water-based inks), place the positive for the second colour over a print of the first colour and tape it in the position where you want it to print. Place the positive under the screen stencil for the second colour and position it so that it directly corresponds to the location of the stencil above. To do this, gently push the screen down with your hands to get a better view and move the paper accordingly underneath. This can be fiddly, but it is worth taking your time as you want all your colours to print in the correct position on each print.

7. Once the print is in the right position, carefully register it with pieces of card as described in Chapter 5 – without moving it, of course! Then print your second colour. Do the same for the following colours until the print is finished.

RIGHT: The different stages for printing this design. *Photos: Megan Randall.*

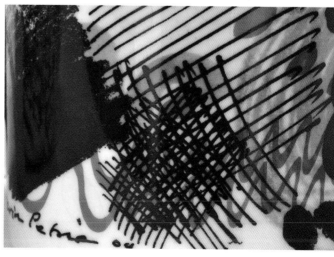

A detail of the fired transfer applied to a mug (see Chapter 5 on how to transfer the print and firing schedules).

CASE STUDY: Robert Dawson – architecture and illusion

Robert Dawson works in an overlap between ceramics, design and fine art and is perhaps best-known for his reinterpretations of traditional ceramics motifs and designs like the famous blue and white 'Willow Pattern'. This work, often applied to plates, offers the viewer details, different perspectives or distorted views of familiar patterns. This gentle subversion reinvigorates these patterns by presenting them in a new way.

Dawson has also taken his approach into the realm of architectural ceramics through his tile pieces. Again these often take traditional design as a starting point, but the introduction of perspective into the two-dimensional surfaces of tiles creates the illusion of three-dimensional spaces.

Play On, Robert Dawson, 2007. 2.5 x 28m (8 x 92ft), print on ceramic tiles. Art House, Wakefield, UK. *Photo: Jeremy Phillips.*

Robert Dawson cutting out transfers and applying them to tiles to create *Play On*. *Photo: Jeremy Phillips.*

BELOW: *Shallow Bond*, diptych, Robert Dawson, 2008. Each panel 30 x 90cm (11¾ x 35½in); overall dimensions 70 x 90cm (27½ x 35½in), print on ceramic wall tiles, image of trowelled tiling adhesive. *Photo: courtesy of the artist.*

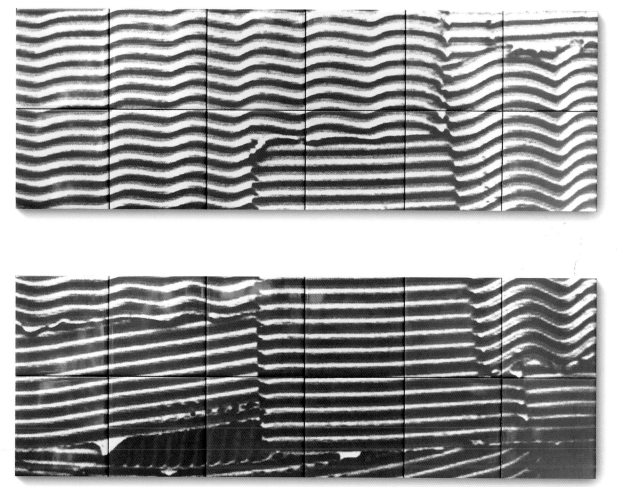

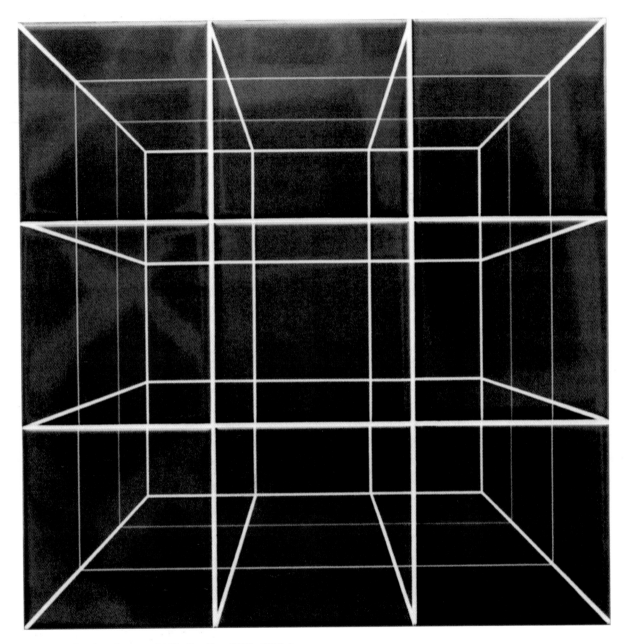

Three Cubed, Robert Dawson, 2005. 45 x 45cm (17¾ x 17¾in),
print on ceramic tiles. *Photo: courtesy of the artist.*

Exterior–Interior, Andrew Livingstone, 2005. 27 x 18 x 4cm (10½ x 7 x 1.5in), slipcast earthenware with transfers. *Collection of the Arts Council Northern Ireland, Photo: courtesy of the artist.*

CASE STUDY: Andrew Livingstone – politics and subversion

Andrew Livingstone's work, which might be termed 'conceptual ceramics', often includes installation, new media, and non-ceramic elements. Transfer printing is important in some pieces, as it allows the transmission of photographic imagery to a new 'location' on a piece of ceramics, thus creating a new layer of meaning. For example, in a series of works exploring the politics of Northern Ireland, the ubiquity of blue and white decoration used to suggest the familiar and domestic is set against the more public face of political agendas. *Exterior–Interior* consists of cast ceramic elements that reference the gable ends of houses. One side is printed with images of the region's political murals in blue and white, while the other is printed with images of wallpaper, again in blue and white. In this work the two prints combine the personal and political.

CASE STUDY: Rob Kesseler – art, science and nature

Ceramics have long been carriers of floral and foliate images, but Rob Kesseler takes these ideas further to embrace science, installation and performance. Collaborating with botanical scientists at Kew Gardens he uses scanning electron microscopes to reveal a hidden world of plants beyond the scope of the human eye. Magnified up to 5,000 times, images of pollen and seeds reveal decorative patterns and complex structures. Working in the fertile territory where fine art, craft and design overlap, printed ceramics often form a cornerstone of his work, translating the scientific images into economic but powerfully exotic botanical icons that sit within but also extend the tradition of natural ornamentation in ceramics. Through installation and performative presentations the work also offers alternative modes of access which challenge viewers' expectations.

Bouquet – Cornflower, Rob Kesseler, 2006. 27 x 13 x 13cm (10½ x 5 x 5in), bone china with gold and enamel printed decals. *Produced in collaboration with Royal Crown Derby. Photo: courtesy of the artist.*

Harvest, Rob Kesseler, 2001, recreated 2007. 60 x 800 x 800cm (24 x 315 x 315in), dried wheat, bone-china plates with gold and enamel prints of pollen. *Installation: The Hub, National Centre for Craft & Design. Photo: courtesy of the artist.*

CASE STUDY: Ceramic print workshops

Rudi Bastiaans (Print Department) and Erik Kok (Ceramics Department) both teach at the AKI/ArtEZ Academy of Visual Arts and Design in Enschede, the Netherlands. The two of them worked with me during a transfer-printing workshop at the University of the West of England in the late 1990s, and have gone on to teach inspirational workshops around the world (with Marcel Vos and Willem Boom).

Their 'Frankenstein' projects bring together photography, image-making, collage, collaboration, computer work, screenprinting, ceramics and site-specific artwork in one project. They work with groups of students to develop new 'Frankenstein's monsters' to be printed onto tiles. They actually screenprint onto the tiles directly rather than using transfers, but I think their innovative project is worth including in this book as the idea could be adapted for transfer printing. The images below are from workshops at Gray's School of Art in Aberdeen, Scotland and Crawford School of Art in Cork, Ireland.

Developing the 'monster' with the positives.

Students are divided into groups and have to plan their new 'monster' design. Photographs are then taken of 'body parts'. The photographs are converted to greyscale and manipulated in Photoshop to form a collage of the monster. This is printed onto normal A4 (210 × 297mm) paper. These sheets are taped together to form a 'life-size monster'. This is then photocopied onto acetate to form positives. These positives are then positioned on the tiles. Separate positives are taped together once the design is completed. These can then be cut into sections to fit the screens. Once the screens are exposed the tiles are direct-printed with on-glaze enamel. After firing overnight the image is reassembled using the positives as a guide.

Rudi and Erik exploited the 'communal' properties of tableware in a project for the AKI Festival where students designed and printed their own transfers. These were then applied to tableware which was used for dinner in a restaurant.

ABOVE AND RIGHT: The artwork can then be viewed in different contexts or even permanently installed.

ABOVE: This image from the International Printing Symposium in Bentlage, Germany shows the great potential of transfers for collage directly onto ceramics to create new narratives. *Photo: Rudi Bastiaans and Erik Kok.*

8

Integrating form and image: one artist's approach

This chapter describes the research of Steve Brown, who has developed innovative approaches to transferring imagery to clay forms, so as to achieve a greater integrity of form and image.

Artist Steve Brown from the UK worked for 14 years in industry as a textile printer. The great experience that he gained from this has underpinned an artistic practice focused on ceramics, but also including glass, which he embarked upon some ten years ago. During this period, he has pursued an exploration of the relationship between the printed image and the form on which it is printed, in an attempt to achieve greater integrity between the two. As transfers are printed onto paper and then applied later to finished, usually glazed forms, there can be a disconnection between the two, with the printed image appearing to 'float' on the object. Of course, this can be used to great effect, and there is a long history of ceramics forming a 'canvas' for painted or printed images that add meaning to the host form.

In contrast, Brown has developed strategies for 'in-mould' decoration that combine the printing with the actual making process of the object. This has evolved over three key stages of development represented by bodies of artworks with increasing complexity. In the first group of works, the images were pressed into moulds. Then, while Brown was studying for an MA, they were actually screenprinted into plaster moulds before the application of clay. Finally, as part of his Ph.D. research, the moulds are still screenprinted but have become flexible fabric, allowing for the integrated manipulation of image and form.

All these systems, outlined below, can be described as transfer techniques in that the printed images are transferred to the clay via a substrate – initially from paper, as is the convention in industry, then from plaster moulds and finally from fabric. The latter approach, a more concerted reappraisal of an appropriate substrate for image transfer, has led Brown back to textiles, the material of his previous career.

Metal dust transfer printing using paper clay

Moreau, Steve Brown, 2009. 50 x 40 x 30cm (19¾ x 15¾ x 11¾in), porcelain with underglaze 'flexible in-mould' transfer printing. *Photo: courtesy of the artist.*

Ghosts in the Machine was the first main project in which Brown developed a personal technique for image transfer. Revising the traditional copperplate method in which viscous ink is printed onto tissue paper, he developed an approach of dusting the print with a mixture of coarse metal and brick dust gathered from industrial sites. Pouring paper-clay slip over the paper encapsulates the image, and when it dries to a plastic state the panels can be collaged together by press-moulding into plaster

mould formers. The metals are affected by the water content in the clay and begin to rust, then firing causes further reactions to occur. Any tissue paper can be used, but it needs to be supported by spray-mounting it onto a stronger waterproof substrate such as an artificial vellum paper; this is peeled away after the slip is poured onto it and firming up has taken place, and before press-moulding.

Brown says of this work that it 'references industry and is an expression of my years working in this environment. Cuttings from health and safety documents are spliced together with engraved imagery depicting workers from the period of the Industrial Revolution. The personnel shown are often women and children working in harsh conditions and the metal print method has been used to express this through reactive print imagery.' Although Brown gathered his own pigments, you could also use underglaze colours or oxides from a ceramic supplier.

'In-mould' stoneware and underglaze transfer printing

For the Architectural Series: 'Hubris', 'Remnant', 'Fragment', Brown worked with a new process using flat-cast plaster bats that he cut up and put together to form moulds. The plaster was worked on as if it were wood, by cutting and drilling into it. Scanning the flat mould sides into the computer, he then used computer aided design (CAD) software to 'fit' imagery to the shapes. The innovation was that Brown screenprinted underglaze colours directly onto the mould sides rather than onto paper. He then pressed clay into the inside of the constructions. As the clay dried it took the image from the plaster and transferred it to the clay surface.

The flat geometry of these formers suggests architectural surfaces, and the

resultant forms are enhanced with decorative imagery that reflects these associations. Arches were photographed from life and printed onto curving mould sides. Making the work in smaller segments and using a wash of porcelain that cracks or shows the clay beneath suggests ageing, memory or perhaps an archaeological architectural find.

Greenware underglaze transfer printing

The aim of Brown's studies 'Transition 1' and 'Awkward Transition' was to work with the conventions of applying decoration to preconceived form. Brown was interested in producing a system for multicolour underglaze printing onto greenware, allowing the surface to continue to be worked on or into. Over the period of exploration the forms became increasingly awkward, in order to create situations where the decoration had to be stretched and manipulated.

These studies were undertaken in order to fully experience and understand the conventional process of analysing existing forms as is the convention within the ceramics industry. Underglaze transfer printing has traditionally been printed onto potter's tissue and applied to biscuit ware as a single colour; multicolour printing is generally prohibitively complicated. Brown's aims were to produce screenprinted underglaze prints that would attach to leatherhard clay and be fired to a mature

Plaster bat mould after screenprinting. Note that each piece is printed separately and then the mould reassembled before the clay is added.
Photo courtesy of the artist.

Underglaze screenprinted transfers made by Steve Brown to fit the form behind. He used a laser level to analyze the form to help design the fit of the transfers. *Photo courtesy of the artist.*

Transition 1, Steve Brown.
Photo courtesy of the artist.

temperature without the glaze taking on a 'matt, unglazed aesthetic'. This also allowed the clay to be carved into or rubbed back around the transfer before it is fired.

The other aim was to take the unwrapped panels from the analysis and scan them into the computer in order to fit the patterns to the form. All the patterns are based on the simple pixel square, which has been manipulated to relate to and articulate the inflection of the forms.

Porcelain and underglaze 'flexible in-mould' transfer printing

Brown created a body of work, which included *Arlecchino the Brave*, *Mandrake I* and *Mandrake II – Double Tuber*, *Moreau*, *Taxonomy series – (After Haeckel) I, II and III*; Brown's aim for this body of work was to redefine transfer printing for the studio ceramicist. Historically, transfers were developed in industry as a means toward specific ends and to address the limitations and requirements of mass production often for tableware or vessels. In relation to this, he says, 'these issues do not apply in the context of my studio practice so I took the process back to its essence in order to fulfil my own requirements. The premise for the relationship between surface image and form within the ceramic discipline predominantly sees the form defined first with the surface dealt with separately. I required a system where this relationship is integrated, and form and surface print can be manipulated together as one.'

Researching different substrates, inks and mediums he adapted the notion of the transfer. Flexible lycra shapes are screenprinted with underglaze colours (mixed with water-based textile medium) and stitched together to create garment-like moulds into which clay can be introduced. These forms can be flexed and manipulated, taking the flat print into three dimensions. Upon firing, the substrate burns away, transferring the image onto the surface of the ceramic work.

Design for *Arlecchino the Brave*. Steve Brown first develops an idea for a piece on the computer. This forms the basis for both the printed decoration and the form.

Arlecchino the Brave, Steve Brown, 2009. 50 x 40 x 30cm (19.68 x 15.74 x 11.81in), porcelain with underglaze flexible in-mould transfer printing. *Photos: courtesy of the artist.*

The pieces Brown has made use surface decoration and imagery to suggest the character of the work. For example, the famous diamonds from the costume of Arlecchino, the most popular of the *zanni* or comic servant characters from the Italian Commedia dell'arte, is referenced in *Arlecchino the Brave*; whilst spiralling, bark-like, antler-like surface imagery is used in *Mandrake I* and *Mandrake II – Double Tuber*.

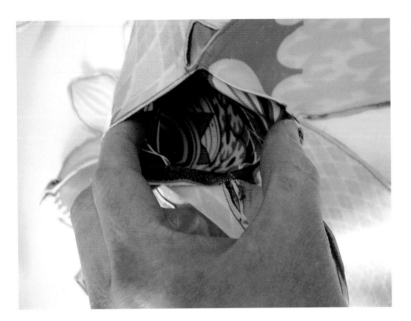

LEFT: Prints on the inside of the Arlecchino mould.

BELOW: 'Moreau' 2. Once the mould is filled with clay it can be manipulated into extreme forms.

ABOVE: Prints for the *Taxonomy* series – empty and full moulds. The printed fabric shapes are then sewn together to form moulds for clay.

ABOVE RIGHT: Injecting clay into the mould. Note: small elements were injected and are solid, while larger elements were press-moulded and are hollow.

RIGHT: *Taxonomy* series (detail), Steve Brown, 2009. Various dimensions, porcelain with underglaze flexible in-mould transfer printing. *Photos: courtesy of the artist.*

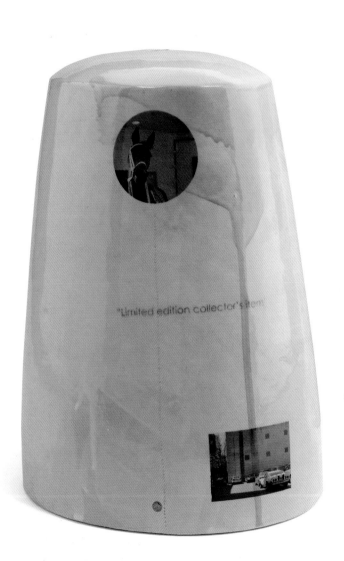

"Limited edition collector's item"

"Limited edition collector's item"

9 Digital transfers

Digital transfer printing machines offer quick and accessible transfer prints. This chapter considers the strengths and weaknesses of these methods.

Digital printing systems use a converted laser printer that prints special enamel colours rather than conventional toners. Using the four process colours of cyan, magenta, yellow and black, the system prints onto an A3 size (15¾ x 11in) ceramic waterslide transfer/ decal paper which is then laminated onto a patented covercoat film. Transfers can then be cut out and applied to ceramics and glass in the same manner as a screenprinted transfer, then fired on to form a permanent bond. The fired images do not fade in sunlight and are dishwasher-safe. This system has been used in large-scale murals on tiles, tableware and photographic memorial plaques. The method is especially useful for colour photographic images, which are difficult to print by screenprinting.

The production of digital prints is very straightforward. An image is scanned or designed on the computer and then sent to a digital ceramic transfer printing company via CD, DVD, ZIP disk or REV disk. The artwork must be in the correct size as the sheet size is A3, giving a print area of under 40 x 28cm (15¾ x 11in). For larger-scale projects, the image can be broken down into sections and printed across several sheets. The image should usually be 300dpi in one of the following formats: Photoshop, Illustrator, Indesign, Acrobat, TIFF, EPS, JPG or PDF.

The design is then printed by the company in the same way as any other computer printer. The cost is about £15 in the UK (approx. $24) per A3 sheet at the time of writing. Ceramic or glass colour is printed onto waterslide paper without any covercoat applied. The company then uses a specially prepared sheet that has a covercoat pre-printed onto a heat-release paper. Both the print paper and cover paper are passed through a laminator. This transfers the covercoat onto the waterslide sheet. Once complete the transfer is sent back to you and simply needs to be cut out and transferred. An order can usually be turned around within a week, making this a cheap and quick way of developing ideas.

Another obvious advantage of this method is that you do not need your own printing equipment and do not have to spend the time learning other print methods. The method is also economic for small numbers of prints. There are, however, some limitations that need to be considered. It is not possible to print white due to the four-colour nature of the process. For ceramic printing this is rarely a problem as the image is usually applied to white-glazed ceramic objects, so it is just like printing four-colour images onto paper. But when applying the digital prints to transparent glass, the areas that should appear white in the image will remain clear.

A second issue is that once fired, especially onto glass, the covercoat layer can

Philadelphia Salt & Pepper Set, Fiona Thompson, 2008. Each 36 x 18cm (14 x 7in), ceramic (earthenware) with digital transfers. *Photo: John McKenzie.*

leave a visible residue. This can be alleviated by cutting out the transfer as close to the image as possible. A third difficulty is that the transfers are only available up to a maximum size of A3. Having said that, larger-scale pieces can be made by 'tiling' images together.

The obvious advantage of screenprint is that it delivers a much thicker deposit of ink and so stronger colours can be achieved. However, printing four-colour screenprints is tricky, so the digital option is much easier. As one would expect, digital prints can look rather like a computer print with halftone dots, which some might feel looks 'cheap'. There can also be some discrepancy of colour once printed. Note that non-firing decals are also available, but these do not form a permanent bond to the ceramic.

My personal view is that it is better not to be seduced by the ease of digital transfers but instead to choose the most appropriate method for the idea. However, in the hands of thoughtful artists digital transfers have great potential, especially when combined with other decoration techniques (see Charlotte Hodes's work in Chapter 3).

CASE STUDY: Alice Mara – humour and everyday experience

Alice Mara is a great advocate of digital transfers and exploits their potential with great flair and wit. She is familiar with a range of ceramic techniques, including screenprinted transfers, but while studying for her master's degree she 'discovered the fantastic process of the digital transfer system, which made it possible to easily develop and actualise my ideas. I could have a "one off" idea and print it digitally with an instant outcome.'

Mara says of her work, 'the concepts and imagery of the work are simple and attractive,

Em's Window Box, Alice Mara. 8 x 6 x 3cm ($3^{1}/_{8}$ x $2^{3}/_{8}$ x $1^{1}/_{8}$in), digitally printed bone china. Of this piece Mara says, 'Looking at and contemplating the white undecorated boxes I wanted the concept of inside and outside to be the crucial factor of the object. The window box was a great way to illustrate this point of inside/outside – using the whole object to create a narrative of everyday life. *Photo: Sonia Read.*

with some layers of humour and everyday experience. I like my work to be easily read by its audience.' The success of her work lies, I believe, in the care and attention she takes in developing the imagery. This includes the taking of her own photographs and their subsequent manipulation on the computer, as well as the selection of appropriate objects to decorate.

Shopping Trolley Cake Stand, Alice Mara. Ht: 34cm (13½in); small plate dia: 17cm (6¾in), medium plate dia: 20cm (8in), large plate dia: 27cm (10¾in), digitally printed bone china. The cake stand imagery of 'endless' shopping trolleys represents an idea of consumerism, greed and abundance which Mara associated with cakes, sweets, and parties and over-indulgence – emphasised by the 'gold garish finish' of the stand. *Photo: courtesy of the artist.*

CASE STUDY: Fiona Thompson – ceramics, travel and tourism

Fiona Thompson is an MPhil graduate of the University of Sunderland, and the theme of her current work is the use of objects and images to evoke memories of a visit to a particular place, particularly the photographic image and the souvenir. Taking this as a starting point, the forms developed are a direct reference to the souvenir. These include the salt & pepper set, the tea tray and the jug. Digital photographs are manipulated in Photoshop, with alterations or distortions made to shape, colour and tone. The commercially made transfers mean that full colour photographic imagery can be used, with the ability to reproduce, repeat and alter relevant images. In an age where digital photographs are accessible, easy to reproduce and disseminate, but transient and often impermanent, there is something particularly appealing for Thompson about rendering them permanent in ceramic.

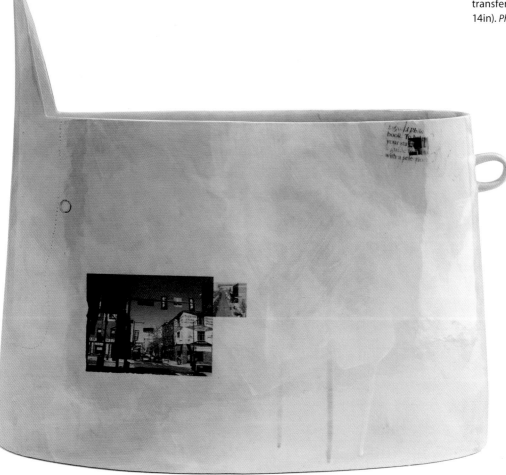

Old City Jug, Fiona Thompson, 2008. Ceramic (earthenware) with digital transfers, 43 x 36cm (17 x 14in). *Photo: John McKenzie.*

CASE STUDY: Robert Winter – giving a voice through clay

Robert Winter has successfully combined his passion for ceramics and his experiences drawn from community work with diverse groups to give a 'voice', through ceramics, to those often denied the opportunity for expression. Robert developed this work by asking people a series of open-ended questions such as 'What is your favorite memory?' or 'Tell me something about yourself that others don't know'. Participants also chose images that were important to them.

Thrown porcelain lidded jars, Robert Winter, 2006. Ht: 35cm (13¾in), porcelain with digital and screenprinted transfers. *Photo: courtesy of the artist.*

CASE STUDY: Claire Turner – text, image and subversion

Claire Turner, a BA student at the University of Sunderland, says of this work, *Isn't it Ironic* is a social comment showing the dichotomy between popular culture and people's reality. I have used popular song lyrics, which layered with an image, create a totally different meaning. I like the idea that the viewer would see the image or the wording first and not realise that the two are combined.'

Isn't it Ironic, Claire Turner, 2009. 10 x 10cm (4 x 4in), slipcast porcelain with digital prints. *Photo: David Williams.*

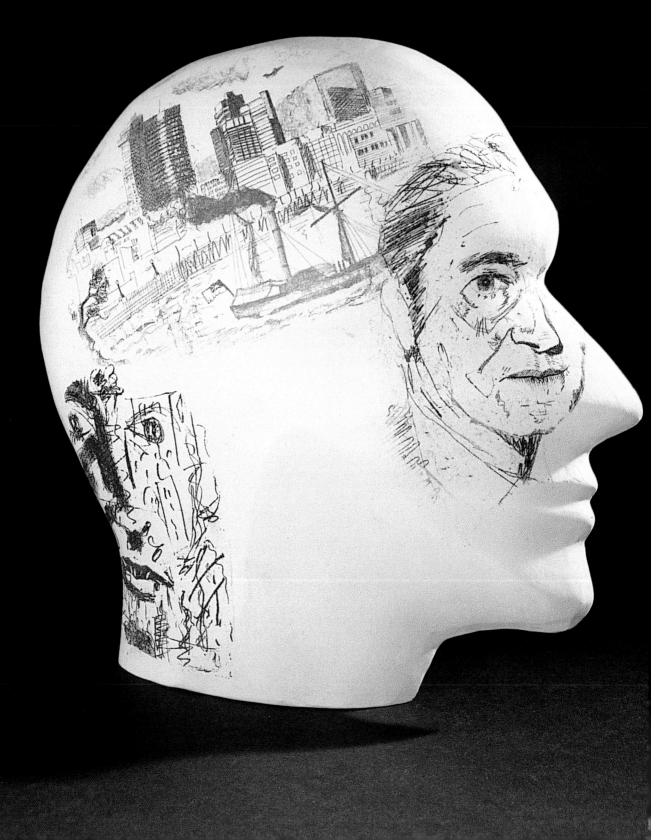

10

Revisiting early transfer printing methods

Early methods of transfer printing offer a distinct aesthetic. This chapter shows how to transfer etchings onto ceramics and describes research using photopolymer plates.

Few artists have exploited the true potential of early transfer-printing methods, which are not as predicable as, say, screenprint but do offer a specific aesthetic (see Chapter 3) that some might wish to exploit. I offer this overview of some of these early methods in the hope that the process can be advanced further. Unfortunately, at the time of writing potter's tissue, needed for these processes, is hard to come by. Despite this, I still felt it was useful to offer this chapter in the hope that a reliable supply, or an alternative, can be located.

I worked on an Arts and Humanities Research Board-funded project at the Centre for Fine Print Research at the University of the West of England (UWE) to explore the potential of photopolymer-plate printing combined with tissue printing. This work was built upon by researcher Tom Sowden at UWE, who has kindly contributed his findings later in the chapter.

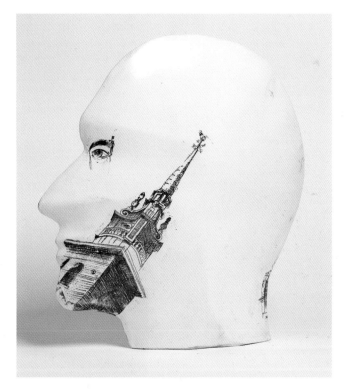

LEFT AND RIGHT: *Head*, Kevin Petrie, *c.* 1994. Ht: approx. 35cm (13¾in), under-glaze tissue print on ceramic. *Photos: (left) courtesy of the artist, and (right) David Williams.*

Printing from an etching plate onto biscuit ware via potter's tissue

This book is too short to cover the process of etching, so I am assuming that a plate has already been prepared.

1 Mix medium copperplate oil (from print-making suppliers) with appropriate colour (cobalt oxide or black glaze stain work well). Add as much colour as possible while retaining a workable consistency of ink. 5g of oil to 10g of colour is a good guide, and will produce enough to print several small plates. Add powder to oil gradually and mix well. Heating the ink slightly on a hotplate aids mixing.

2 Place the etched plate on a hotplate and apply ink with a stiff piece of card. The heat will loosen the ink and make application easier. Remove from the heat and work the ink into the plate using a ball of scrim (tarlatan). A circular downward movement ensures the ink fills all the etched areas. Take a clean piece of scrim and wipe any excess ink from the surface of the plate, without removing ink from the grooves. A slight residue of ink will not transfer.

3 Paint a solution of half soft soap and half water sparingly on one side of a piece of potter's tissue. This helps stop the tissue sticking to the plate and keeps it flexible while the image is being transferred. Place the inked plate onto a hotplate again to loosen the ink slightly. Then place on the bed of the etching press. Lay potter's

BELOW LEFT: Applying ink to the plate.

BELOW RIGHT: Removing the printed tissue.

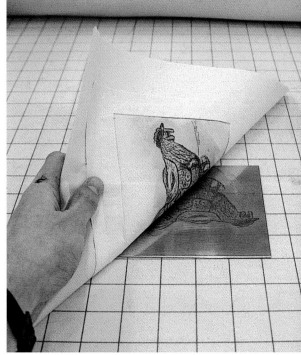

ABOVE LEFT: The transferred print.

ABOVE RIGHT: *Nan's Teapot*, Kevin Petrie, fired and glazed tile. *Photo: David Williams.*

tissue, sized side up, onto the plate. Cover with a damp piece of paper, then a clean piece of paper to protect the blankets on the press. Run through the press and remove the printed potter's tissue from the other pieces of paper.

4   Place the printed tissue, image side down, on bisque ceramic and, working from the centre outwards, smooth the image down with your fingers so that the tacky ink adheres to the ceramic. Paint soft soap size onto the tissue, again from the centre outwards, so that it lies flat. Take a rubber kidney and firmly rub the back of the tissue to offset the image onto the ceramic. Lift a corner of the transfer to check that the image has transferred. If it hasn't, rub some more. Remove the tissue from the ceramic. The piece will require a 'hardening on' (680–700°C/1256–1292°F) firing before glazing, to remove the printing medium.

Printing from a flexography plate onto ceramic via potter's tissue

Photopolymer-plate printing, sometimes known as flexography or solarplate printing, is a commercial process used for packaging. A photopolymer plate consists of a backing sheet, usually steel or aluminum, a photosensitive polymer layer that hardens when exposed to UV light, and an adhesive layer that bonds the polymer to the backing plate and prevents light reflection. The surface of the polymer is protected with a plastic cover film which must be removed before use.

Tile, Tom Sowden. Underglaze-printed via potter's tissue from relief photopolymer plate. *Photo: courtesy of the artist.*

The plate is exposed to ultraviolet light and developed in warm water. This produces a printing plate with a relief surface of between 1 and 4mm. Conventionally, ink is applied to the plate either on the uppermost surface (relief printing) or in the 'depressed' areas (intaglio printing), and the image is printed onto paper or card, but photopolymer plates also have interesting potential for printing onto other substrates like ceramics and glass (see my book *Glass and Print*, also published by A&C Black).

As mentioned above, I worked on a research project at UWE which explored the potential for using photopolymer plates to produce four-colour (CMYK) prints, rather like the 19th-century four-colour prints used on pot lids. For this project we used intaglio photopolymer plates, inking them up and printing them like you would an etching plate in a similar manner to that described above. Later, Tom Sowden achieved some success in transfer-printing underglaze pigment onto biscuit-fired tiles using relief photopolymer plates and potter's tissue. The process is also suitable for both single-colour printing and four-colour process printing (CMYK).

Having made your photopolymer plate (see the step-by-step guide on pp.113–4) mix underglaze pigment with a lithographic printing medium (Vanson Rubber Base

Plus or PANTONE Transparent White). Roll ink out onto the glass first, then onto the photopolymer plate. Several passes will build a sufficient layer of ink. Cut a piece of potter's tissue slightly larger than the plate, allowing for any movement that occurs when placing it. Then place the potter's tissue on the inked surface, gently rubbing the back of the tissue with your finger to pick up the ink.

Carefully peel the tissue off the plate and apply face down to the ceramic. If you are printing using a four-colour printing process (CMYK), line up the tissue with the registration marks of previously printed colours. Rub the back of the tissue to offset the image. When ink starts to adhere to the surface, wet your fingers with warm water and plenty of soap (plain white bar soap) and continue to rub back, so as to release the remainder of the ink from the tissue. Once the whole image area has been soaked in soap and water, peel away the tissue and the printed image should be left behind.

On ceramics, Tom had the most success using underglaze on biscuit-fired tiles, which were subsequently glazed. He also managed to achieve good-quality single-colour results, using the same process, on enamel on metal with a compound surface (by mixing on-glaze pigments rather than underglaze).

Producing relief photopolymer plates for ceramic transfer printing

1 An image is produced on the computer. This can be a vector-based image (as in the example below) or a photographic image. Output onto transparent film with a laser printer. Relief plates need to be printed as reversed negatives of the original and, if photographic, with halftone screen, 50 lines per inch (lpi). Intaglio plates need a positive image.

A photopolymer being washed out. Note the areas of metal that are revealed towards the end of the process. *Photo: Tom Sowden and David Sully.*

2 Cut the photopolymer plate to size and place it on a UV exposure unit, emulsion side up. Remove the protective plastic coating, then place the printed film positive on top of the plate, emulsion side down, secured under glass with a vacuum. Next set the exposure time. This will vary according to the exposure unit used. In the example illustrated an exposure time of 35 seconds was used.

3 The plate is then put into warm water and rubbed with a soft sponge or brush. As the plate is washed, the photopolymer in the negative areas, which are black on the film, washes away. Continue washing until all the photopolymer from the negative areas has gone and the metal is exposed. Then the plate is dried and exposed again (covered with protective transparent film) to harden. It can then be printed as described above and transferred to ceramic in a similar manner to that described for transferring an image from an etching plate.

CASE STUDY: Charlotte Hodes at Spode – a collage of traditions

Charlotte Hodes worked on one-off pieces using Spode ware onto which she collaged fragments of tissue-printed transfers from Spode's original engraved copper plates. She worked on the factory floor, combining underglaze colours and using tissue transfers that happened to be in production on the day she produced three unique dinner services.

A Dinner Service (detail), Charlotte Hodes, 2001. Dinner service with four place settings: underglaze tissue transfer from the Spode archive onto Spode white china. *Photo: Peter Abrahams. Courtesy of Marlborough Fine Art.*

CASE STUDY: Megan Randall – distorting traditions

Megan is interested in how images on 'Willow Pattern' ceramics have changed along with the story and how both story and image have distorted over time. In this work, she screenprinted oxides, mixed with glycerin B.P. (glycerol), onto tissue paper (not potter's tissue, just cheap gift-wrap tissue – although she did find the slightly waxed interleaving paper that comes with readymade transfers worked well), and transferred them to leatherhard porcelain, which was then biscuit-fired and glazed.

Distorted Traditions, Megan Randall, 2009. Approx. ht: 5cm (2in), tissue-printed transfers on porcelain.
Photo: courtesy of the artist.

11

Transfer printing and enamel on metal

Vitreous enamel on metal offers a rich alternative surface for the application of prints.

Vitreous enamel is essentially a layer of clear or coloured glass fused to a metal surface (commonly steel, copper, gold or silver) through the action of heat (760–820°C/1400–1508°F). The material applied to the metal is called vitreous enamel, and the process of fusing (bonding glass to metal) is called enamelling. The finished product is called enamel, or glass on metal.

Enamel on metal has many applications including jewellery, boxes (traditionally snuffboxes), plates and mugs (for camping), photographs on gravestones and memorials and large decorated panels such as those seen in underground train stations, as well as for architectural cladding, public-art commissions, household appliances, and signage.

An enamelled surface can be achieved through a range of approaches including painting, spraying, pouring, abrading, and drawing through layers. Enamel on metal differs from glass and ceramics in that layers of enamel can be fused onto pieces of metal in just a few minutes in the kiln. This allows layers of enamel to be rapidly built up through several firings in a short time. Enamel is also very durable and more resistant to extremes of temperature than ceramic or glass. This makes it an ideal material to use outside.

Enamelling is an art form in its own right and is too broad a subject to cover in detail in this book. Below are some notes for using transfers on enamel on metal, prepared by the eminent enamel artist Elizabeth Turrell, Senior Research Fellow in Enamel at the University of the West of England, Bristol. For more information see the Further Reading section at the end of this book.

Using water-based transfers on enamel on metal

Court Date Corsage (brooch), Kathleen Browne, 2006. 21.5 x 16.5 x 1.25cm (8½ x 6½ x ½in), fine silver, sterling silver, rhinestones, vitreous enamel. *Photo: Kathleen Browne.*

Finely ground enamel powders (on-glaze or underglaze) are mixed with a water-based medium and screenprinted onto transfer paper in the same manner as transfers for ceramics. This is applied to the enamel surface in the same way as for ceramics. Digital prints can also be used.

The transferred image is most stable on pre-enamelled steel. On copper, use hard- or medium-fusing enamel as the background enamel for a more stable image (ask the supplier of enamels to advise on this). Soft-fusing enamels can cause the transfer to craze or create a 'soft image' when fired.

Maturing temperatures of enamels are approximately as follows:

Hard: 816–870°C (1500–1600°F)
Medium: 788–816°C (1450–1500°F)
Soft: 760–788°C (1400–1450°F)

Test the transfer on your background colour(s), as each enamel contains different metallic oxides as well as having slightly different maturing temperatures. This affects the way the transfer image settles into the background colour(s) when fired.

Applying the transfer

It is easiest to apply a transfer to a surface that is flat, though it is also possible to apply the transfer to a gently curving surface. It should always be applied to a pre-enamelled surface, i.e. a metal surface that has a layer of enamel already fused to it. The surface should also be cleaned thoroughly to remove any dust or grease before transfer application.

Firing

Do not put transfers that are damp in the kiln; they must be absolutely dry before firing.

If the applied transfer is put in the kiln at the usual temperature for firing enamel, the plastic carrier (or covercoat) will burst into flame and may well damage the image, resulting in loss of areas of the on-glaze image. Place the piece or pieces with the dried transfers on a trivet on a firing mesh and then in a cold kiln (fit in as many pieces as the kiln will hold). Turn the kiln on low and heat slowly to 400–450°C (750–840°F). This will gradually burn off the plastic carrier; but ventilate the area well at this point as the plastic carrier has an unpleasant smell. The transfer will become dark brown in colour. This will gradually fade as the plastic burns off, until the image on the transfer appears clean and clear. At this point the piece can be removed from the kiln and one of the following procedures can be followed.

If you have access to two kilns you can burn off the covercoat in one kiln and have the second kiln at approximately 760°C (1400°F). When you have reached 400–450°C (750–840°F), remove from the first kiln and place in the second kiln immediately – for between one and three minutes depending on the size of piece and the temperature of the kiln.

If the use of two kilns is not possible, remove the piece at 400–450°C (750–840°F) and carefully put it in a draught-free place. Obviously at this point the pigment has not bonded to the pre-enamelled surface and can easily move or flake. Turn up the kiln to full and fire to between 700°C (1290°F) and 760°C (1400°F).

The following burning-out method can be achieved in a reasonably short time and seems to work for most of the transfers:

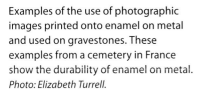

Examples of the use of photographic images printed onto enamel on metal and used on gravestones. These examples from a cemetery in France show the durability of enamel on metal. *Photo: Elizabeth Turrell.*

Applying enamel transfers. In this piece from her *Catching Numbers* series, Elizabeth Turrell is positioning a transfer of numbers onto a piece of enamelled steel. The enamel surface consists of a layer of poured white wet-process enamel that has been drawn through to reveal the fired black enamel grip coat beneath (seen as black lines), and then fired. Thin layers of wet-process enamel have then been painted over using a masking-tape stencil, and fired to create defined translucent layers. The bold red numbers will provide a graphic counterpoint to this surface once applied and fired on.

1 Place the enamel in the kiln and turn to low.

2 Allow the temperature to rise over about an hour and a half to 175–200°C (350–390°F).

3 Then the temperature should rise to 315–345°C (600–650°F). The transfer will look dirty brown at this stage.

4 Then allow the temperature to get to 370–400°C (700–750°F).

5 Check that the entire brown residue has burnt away before removing the enamel from the kiln.

6 Fire at 700–760°C (1290–1400°F). There is physically very little pigment on a transfer, so it only needs a brief firing.

7 If you need to burn off other pieces, turn off the kiln, as it doesn't need to be cold to begin the process again. For example, if the kiln is at 400°C (750°F) leave the door ajar for a few minutes and then you can place another piece in the kiln straightaway; the piece and the mesh will lower the temperature further. After you have left the door open until the temperature drops to 200°C (390°F), you can close the door for a few minutes, then switch on the kiln and proceed as above.

If you cannot take this much time to burn off the plastic carrier, you can increase the speed of burning off by moving the pieces in and out of the kiln for a few seconds at a time, until the covercoat has burned away and the transfer looks clean. This works

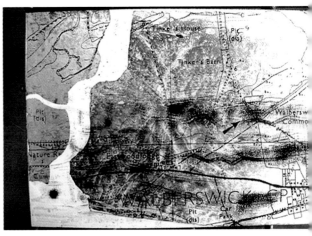

especially well on small pieces. More complicated and four-coloured transfers, especially those that are digitally printed, need to be burnt off slowly.

Sometimes breaking up and curling of the surface can occur with a four-coloured (CMYK) transfer and those with dense colours. If this happens, try firing the enamel from cold (see above).

Some general points on firing transfers and methods

To achieve 'true' colour from transfers, adequate venting is vital. Problems relating to poor venting include poor colour development and a cloudy or hazy appearance.

Transfers that are underfired or overfired may develop the following: faded colours (overfired), colour shift (underfired), transfers rubbing off (underfired), dull appearance of the surface (underfired).

Regarding firing range it is advisable to test-fire transfers, as the colour can be easily affected by the amount of heat work done. Make several different firings and then select the best results as your firing programme.

The following approaches work well:

- Transparent enamels painted or poured over the fired transfers and then fused.
- Washes of on-glaze colours painted over the fired transfers.
- Soft or misty images can be created by layers of fired decals separated by layers of thin, sifted or wet-process (poured) enamel fired over the transfer and then abraded with a stone lightly to reveal the imagery.
- Building up several layers of transfers, firing in between each layer.
- Using lustre over the fired transfers.

These instructions aim to achieve a perfect surface for the transfers. However, it is open to subversion by artists! And that is just a starting point. Once you have found out the most successful procedure for your work – experiment!

ABOVE LEFT: Transfers on a flat sheet of metal placed in the kiln before the door is closed.

ABOVE: A transfer on metal in the process of being burned off. Note the brown area, which is the partly burned-off covercoat layer. When this brown colour has disappeared the covercoat has fully burned away.
Photos: Elizabeth Turrell.

CASE STUDY: Elizabeth Turrell – a personal perspective on enamel and print

Elizabeth says of her work, 'I am compelled to make markers and memorials, both to remember individuals and to mark conflicts. Maybe, in today's world, I am making talismans. I find transfers suit my working process as I can have a 'bank' of images on hand. I start by working on a piece using direct drawing and painting techniques on the unfired enamel. The work/image is built up using layers of enamel on the metal. Each layer is fused to the subsequent layer of fired enamel. The use of print creates possibilities of repeated images and a formal print quality. The most interesting eventuality for me is the option of developing soft and half-hidden printed images, which almost dissolve into the layers of enamel during successive firings.'

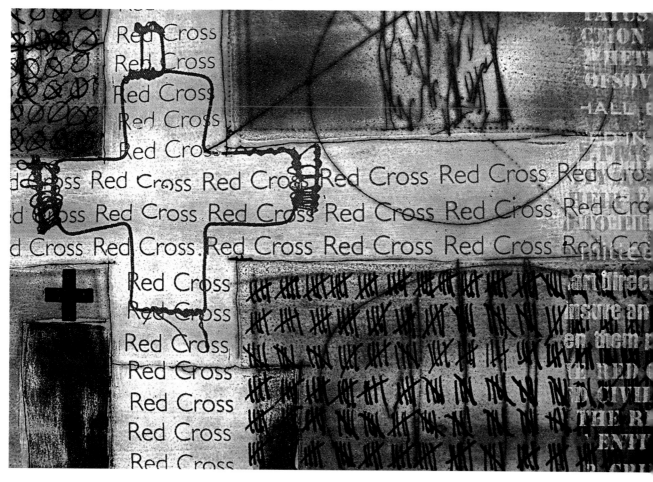

Red Cross (detail), Elizabeth Turrell, from the 'Universal Declaration of Human Rights' series, 2009. 29 x 21 x 1cm (11½ x 8¼ x ⅛in), vitreous enamel on steel, drawn and painted with transfers. *Photo: Elizabeth Turrell.*

CASE STUDY: Kathleen Browne – enamel, transfers and jewellery

US-based artist Kathleen Browne is an excellent exponent of the use of transfers on enamel in jewellery. Her work exploits the historical conventions of miniature portrait enamels, and often draws upon US pulp-fiction magazines of the 1950s.

Worry (necklace), Kathleen Browne, 2004. 54.5 x 18 x 6cm (21½ x 7 x ¼in), fine silver, sterling silver, vitreous enamel. *Photo: Ralph Gabriner.*

CASE STUDY: Matthew Partington – fixing memory

Matthew says of this work, 'My grandmother suffered from dementia. As a result a spoon was 'that thing you eat with'. The enamelled squares refer to the child's game of 'Fifteen', whereby a series of tiles have to be slid into place to make an image. The tiles have been put together so they will never make the complete image of the spoons, and cannot move, reflecting the way in which Emily's dementia meant she could never remember the name for a spoon.'

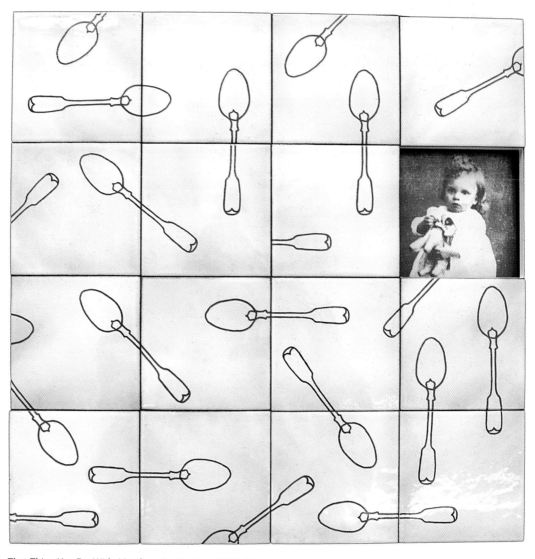

That Thing You Eat With, Matthew Partington, 2007. 40 x 40cm (15¾ x 15¾in) (unframed), 60 x 60cm (23½ x 23½in) (framed), screenprinted enamel transfers on vitreous enamel on copper. *Photo: courtesy of the artist.*

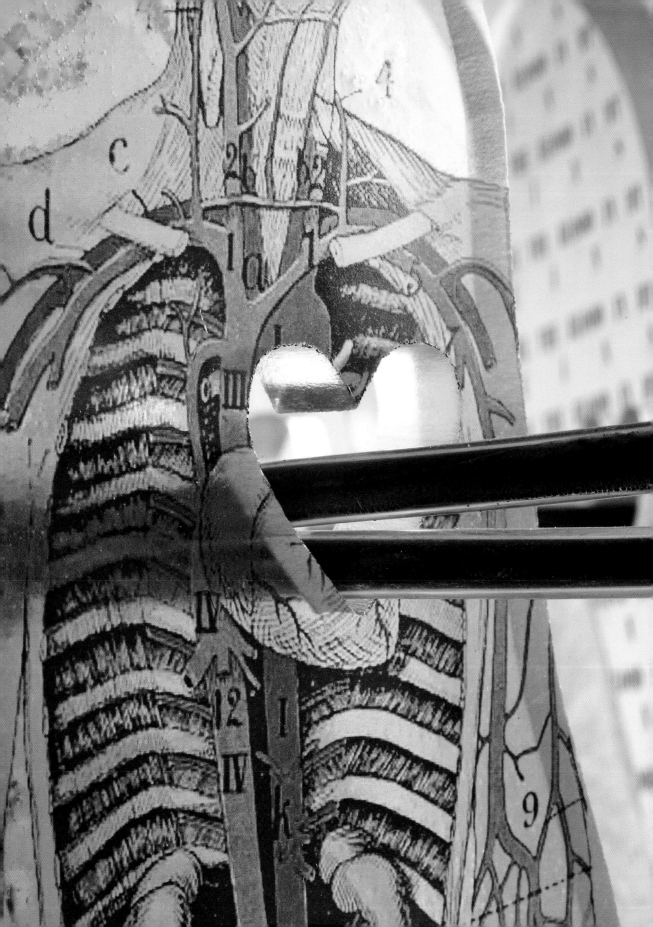

12

Transfers and glass

Glass offers the potential for distortion, transparency and the encasing of imagery inside forms.

Ceramic transfers (i.e. those made with on-glaze enamels) can be used in a number of ways inside glass as opposed to just on the surface. For example, transfers can be applied to a blown glass form and fired in a kiln. This can then be taken from the kiln and a layer of glass gathered over the top from the furnace. It can then be blown into a form, which distorts the images on the transfer. Transfers can also be applied to ingots of casting glass and laid into a casting mould. This is placed in the kiln and heated so that the glass takes on the form of the mould and the transfer moves with the glass to create strange distortions. Flat sheets of glass can also be printed and fused together to create layers of imagery. I covered some of these approaches in my book *Glass and Print*, but can offer some new examples here through case studies that show a range of approaches in this area.

Bloodline, Claire Turner, 2009 (detail, left). 18 x 42 x 1cm (7 x 16½ x ³⁄₈in), water-jet-cut glass with digital transfers, plastic tubing with human blood. A BA student at the University of Sunderland, Claire Turner says of this work, '*Bloodline* represents the connection that naturally occurs between blood and organ donators and their receivers. The idea that people are being saved by a stranger's selfless donations is romantic. I have used unidentifiable shapes as a representation of each individual person. I wanted to use the human blood, imagery and the cut-out hearts in each shape to represent the connection between each person.' *Photos: David Williams*

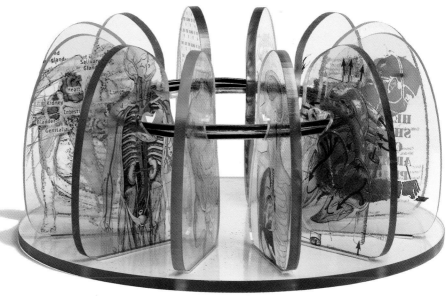

CASE STUDY: Shelley James – imaging the eye

Shelley's time as Artist in Residence at Bristol Eye Hospital opened an incredible window into medical imagery and how perception works, and has formed the basis for a strong body of works combining transfers and glass. The past five years have seen a revolution in our understanding of cell differentiation, through which all the structures of the body are formed and maintained. Stem cells are also referred to as 'pluripotential', meaning that they have the potential to express a wide range of different characteristics, depending on their environment. Recent research suggests that pluripotential cells placed in the deep purple lining at the back of the eye can grow to recreate failing links in the visual pathway.

A series of clear glass 'cells' with printed expressions were encapsulated in hot glass along with deep purple cells, engraved to trap bubbles. The form was then cut and polished by hand.

Pluripotential 1, Shelley James, 2008. 14cm x 14cm (5½ x 5½in), hot glass and ceramic transfer. Blown by Sonja Klingler. Note the printed transfer of the eye inside the glass.

RIGHT: An interesting experimental piece by Shelley James using digital transfers on different layers of glass to create a sense of movement. The transfers are fired onto separate pieces then presented together in a simple frame. *Photos: courtesy of the artist.*

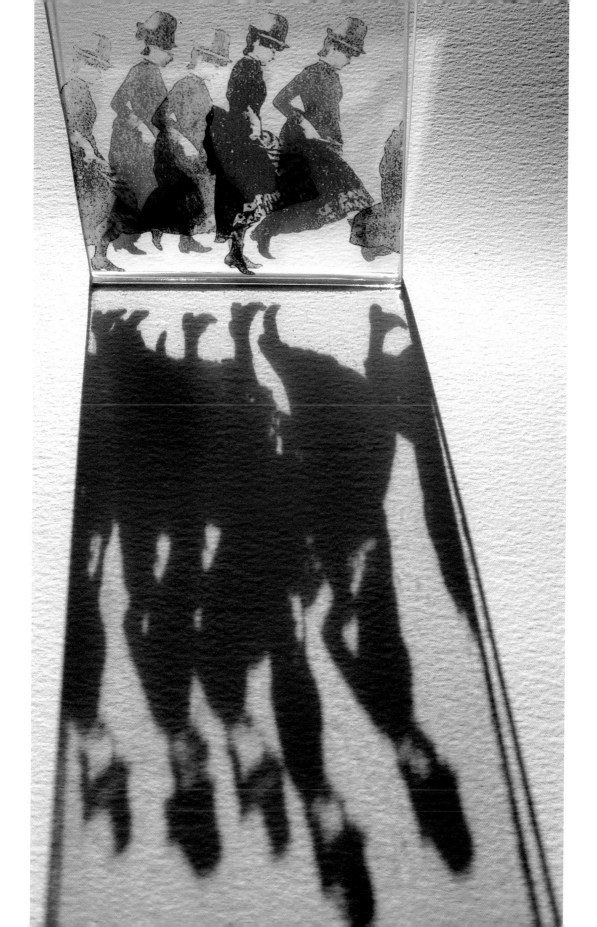

CASE STUDY: Jeffrey Sarmiento – cultural identity: digital transfers in glass

Glass offers the possibility of layering various graphic images while making them simultaneously visible. Jeffrey Sarmiento's work *Fight/Flag* utilises the graphic image 'in' glass form to express his mixed cultural identity. The source images were taken in the United States, where his uncle, a wealthy Filipino immigrant, began to raise and train fighting cocks on his property in a gated suburban community. The clash of contemporary perceptions about animal cruelty against a traditional sport with ritual and social implications in his ancestral homeland inspired this work. After the work was made and exhibited, Sarmiento learned that his uncle had been arrested in a raid on illegally organised cockfights. The rooster wears a leash, a sign that it is being raised for sport. Printed images of a cock's spurs in the colour of the Filipino flag float in layers placed perpendicular to the image of the rooster sitting behind. This work is made by layering pieces of flat glass together and fusing them in the kiln. Grinding and polishing reveals the prints inside. Fighter uses the same cock motif printed onto white glass.

LEFT: *Fight/Flag*, Jeffrey Sarmiento, 2007. 36 x 30 x 3cm (14¼ x 11¾ x 1¼in), screenprinted glass with digital decals (cockerel), fused and polished. *Photo: Kent Rogowski.*

RIGHT: *Fighter*, Jeffrey Sarmiento, 2007. 110 x 80 x 4cm (43 x 31½ x 1½in), blown glass with digital transfers. *Photo: Kent Rogowski.*

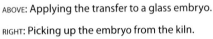

ABOVE: Applying the transfer to a glass embryo.

RIGHT: Picking up the embryo from the kiln.

Photos: Stephen Beardsell.

CASE STUDY: Kathryn Wightman – illusions of textiles: blown glass with screenprinted transfers

Kathryn Wightman has undertaken Ph.D. research at the University of Sunderland into the development of novel print-inspired approaches for use with blown glass. Her style is interesting in that her use of pattern gives a textile-like quality reminiscent of printed fabric. When the imagery is trapped within layers of glass it also assumes an intriguing three-dimensional effect.

To achieve the desired look, Wightman uses screenprinted transfers in combination with other glass techniques such as 'graal' (a method using several layers of coloured glass in which layers are sandblasted through to create patterns) to collage and layer her imagery.

The following images show how Kathryn Wightman collages and layers her imagery within blown glass to create her 'illusions of textiles'.

A green 'embryo' of glass with a white glass overlay is blown. The embryo is taped up to form a stencil of eight evenly spaced vertical stripes. The un-taped part of the embryo is sandblasted to remove the top layer of colour. The remaining tape is removed and the embryo cleaned. The pattern is manipulated using Photoshop to ensure that it fits perfectly onto the white stripes. The image is printed using an

ABOVE LEFT: A gather of glass is taken to cover the embryo and the post. The post is the piece of glass which connects the embryo to the blowing iron. As only a small layer of glass is needed, the excess is run off into a metal bucket.

RIGHT: Kathryn Wightman shaping the form.

Photos: Stephen Beardsell.

underglaze colour (high-temperature black) mixed with solvent-based screenprinting medium. The transfer paper is U-WET.

The transfers are applied. Then the embryo is dried, transferred to a top-loading kiln in the hot shop, and heated up at a temperature of approximately 40°C (72°F) an hour until it reaches 540°C (1000°F). This burns away the covercoat layer, leaving the high-temperature colour. Once this temperature is reached, the embryo is ready to be picked up on a blowing iron and blown and shaped further.

The embryo is then transferred to the glory hole and heated. The transfer is fired on to the embryo during this initial heating in the glory hole, meaning that the colour melts and bonds to the glass. It is evident that firing-on of the transfer has taken place when the appearance of the print changes from matt to gloss.

Once the print is fired, shaping can begin. The most effective method is 'marvering' (shaping the glass on a metal table as opposed to a paper pad), as it tends not to damage the image too much during shaping. Once a suitable shape is achieved, the embryo is cooled slightly in preparation for 'gathering over' with a skin of clear glass to enclose the printed image within the form.

The gather is then shaped using a damp paper pad. Further shaping is undertaken to achieve a teardrop shape. Once the piece is complete, it is transferred to a lehr (a special kiln used specifically for annealing glass) to cool down slowly.

Kathryn has also been investigating the potential of creating thin patterned sheets of glass using a photopolymer plate. She takes a plaster and molochite mould from the relief surface of the plate and presses glass powders into the mould. These are then fused together in the kiln. These thin sheets are then transferred by a roll-up method onto a blown-glass form, and encased in layers of glass. This process creates a transfer that is denser than that of a screenprinted transfer, offering the potential to create much larger blown-glass forms.

BELOW: *Dress Up Your Home!* Kathryn Wightman, 2007. 25cm x 11cm (9¾ x 4¼in). Graal pick-up with sandblasted checks and screenprinted transfers encased in layers of hot glass. *Photo: David Williams*.

RIGHT: Kathryn Wightman. A flat panel of glass made from a polymer plate. This experimental work of translating a panel into hot glass is ongoing. *Photo: Stephen Beardsell*.

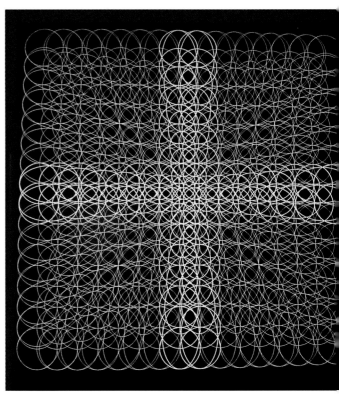

ABOVE: A printed black square is applied to borosilicate tube and smoothed down. It is then etched in the round with a laser and fired on with a torch.

RIGHT: A flat sheet of glass where the transfer has been laser-etched.

Case study: Steve Brown – borosilicate transfer laser etching (in the round)

Borosilicate glass is a type of glass that includes boric oxide, making it resistant to extreme temperatures. This series of studies explored the possibilities of using a laser to etch through readily available single-colour transfer sheets.

This method for working with transfer sheets eliminates the need to print the work oneself. The sheets can be bought in and the laser etching carried out on flat ware and in the round by a specialist bureau. Brown explored the flat samples using black and gold-lustre sheets, with the latter being produced much more cheaply than by the usual route of buying expensive pigments, making a screen and printing it conventionally.

RIGHT: The tube blown into a vessel. *Photos: Steve Brown.*

CASE STUDY: Catherine Labonté – storytelling with glass

Canadian Catherine Labonté is a glass artist inspired by animal behaviour and cartoons. With her 'bell jar' sculptures she is 'telling you a different story every time. But the best way to read them is to let yourself imagine and create your own.' She hopes to create objects that, 'evoke laughter, wonder and lighten the heart'.

She is using screenprinted transfers because, she believes, 'it is the best way to recreate my drawings on glass. With screenprinting I can multiply one drawing so that one tree can become a forest (*Dilemma*). With transfers I can get the fine results that could not be achieved with direct glass painting, which is not as precise.'

Catherine Labonté, *Dilemme/ Dilemma*, 2009. 44.5 x 25.5 x 32cm (17½ x 10 x 12½in), blown glass, kiln-formed glass, cast glass, transfer, glass, wood. *Photo: Catherine Labonté.*

Conclusion

I was surprised while working on this book to find such exciting approaches and enthusiasm for what might at first be thought of as a rather narrow subject. This leaves me feeling optimistic for the future.

As I wrote at the end of my last book, creative vision rather than technique is what leads to successful art and design. I hope that this book can offer both some useful methods and a little inspiration that others might build upon to continue to develop the powerful potential at the overlap between ceramics and printmaking into the new decade.

Kevin Petrie – National Glass Centre, University of Sunderland

Colour Me a Rainbow, Kathryn Wightman, 2006. Hand blown glass incorporating multiple layers of screenprinted transfers.
Photo: David Williams.

Suppliers

Colour ('enamels') – on-glaze, in-glaze, underglaze, glass

Amaco Ceramic Supply
USA
www.amaco.com
Ceramic suppliers including on-glaze/overglaze and underglaze.

Dove Screen Services Ltd
www.dovescreen.co.uk
Screenprint supplies. On-glaze, underglaze and in-glaze colours. Glass colours. Transfer papers.

Ferro Corporate Headquarters
http://www.ferro.com
On-glaze, in-glaze or underglaze systems, decoration media (solvent-based), covercoats and auxiliaries.

Johnson Matthey Colour Technologies
http://www.colour.matthey.com/
World supplier of on-glaze, in-glaze or underglaze systems, media (solvent-based) and covercoats.

Potterycrafts
www.potterycrafts.co.uk
Major supplier of ceramics and glass materials and equipment including kilns.

W. C. Heraeus
http://heraeus-ceramiccolours.com
Major world supplier of ceramics and glass colours, printing media (solvent-based) and covercoats.

Screenprint suppliers

Cadisch Precision Meshes Ltd
Email: info@cadisch.com
Mark Resist 150 microns, Azocol Z1 screen emulsion, new and re-stretched screens, squeegees, mixing cups, parcel tape and sundry items.

Daler Rowney Limited
Fine Art & Graphic Materials
www.daler-rowney.com
System 3 drawing fluid and removable screen block. Also mediums for water-based printing.

Folex AG
www.folex.com
Suppliers of Matt Laser Film (FOLAPROOF LASERFILM/F) for printing positives from the computer or photocopier. Also good for hand-cut stencils.

Fortune & Associates
Tel +44 (0)1934 732 691
or +44 (0)7973 776 019
Email: fortune.eden@virgin.net
Portable exposure units and tabletop screenprinting beds ideal for schools, colleges and artists' studios.

Screenstretch Ltd
http://www.screenstretch.co.uk
Screens, coating troughs, squeegees, frame re-covering service, preparation & screen reclaiming chemicals, emulsions.

Harvest (detail), Rob Kesseler, 2001, recreated 2007. Dried wheat, bone-china plates with gold and enamel prints of pollen, 60 x 800 x 800cm (23½ x 315 x 315in). *Installation: The Hub, National Centre for Craft & Design. Photo: courtesy of the artist.*

Water-based transfer printing

John Purcell Paper
Email: jpp@johnpurcell.net
U-WET transfer paper, T.W. Graphics
water-based ink system, Flat Clear
Base for water-based transfer printing,
TrueGrain.

Open-stock and custom-run transfers

Bailey Decal Ltd
http://www.baileydecal.co.uk
Rubber stamps, open-stock and custom-run transfers.

Digital transfers

Ceramic Digital
www.ceramicdigital.co.uk
High-resolution transfers (up to 300dpi).

ceramic-digital-transfers.co.uk
(trading under Heraldic Pottery Limited)
www.ceramic-digital-transfers.co.uk
Digital and screenprinted transfers.

Ceramital Limited
www.ceramital.com
Integrated transfer printing systems,
consumables, on-demand digital
transfer printing and application.

DecalPaper.com
USA
www.decalpaper.com
Non-firing decals for ceramics and glass.

FotoCeramic
www.fotoceramic.com
Bespoke production of ceramic transfers.

inplainsight art
http://www.inplainsightart.com
Custom imagery permanently fused to
glass and ceramic tiles. Offers 'artist
decal' service.

Whiteware ceramics for decorating

Fegg Hayes Pottery Ltd
www.fegghayespottery.co.uk
Wholesale whiteware for decoration,
special runs to customers' designs,
in-house design, in-house firing,
in-house decoration.

White World UK Ltd
www.whitebonechina.co.uk
White bone china supplier.

Enamel-on-metal suppliers

Allcraft Jewelry Supply Co.
USA
Tel: +1 718 789 2800 (mail order)/800
645 7124 /212 840 1860
Fax: +1 800 645 7125
Latham enamels, Vcella kilns.

Bovano of Cheshire
USA
www.bovano.com
Cristallerie de Saint-Paul and Thompson
enamels.

Cristallerie de Saint-Paul
France
www.emaux-soyer.com
Soyer enamels for copper, silver
and gold.

Ellen Goldman
The Netherlands
www.goldman-enamel.com
Thompson enamels and useful technical
information.

Enamel Emporium
USA
www.enamelemporium.com
Ninomiya enamels, other enamels and
products.

Enamelwork Supply Co.
USA
www.enamelworksupply.com
Enamels, enamel supplies. Schauer and
Ninomiya enamels.

Fred Aldous Ltd
www.fredaldous.co.uk
Enamels, copper blanks, pewter sheet,
copper sheet, kilns and tools.

Thompson Enamels
USA
www.thompsonenamel.com
Thompson enamels/enamelling supplies
and kilns.

W.G. Ball Ltd
www.wgball.com
Manufacturers of lead-free enamels,
ceramic colours and oxides for vitreous
enamels. Also suppliers of kilns, tools,
findings and transfers.

Tissue Printing

Northcote Pottery Supplies
Australia
www.northcotepottery.com.au
Pottery supplies and ready-printed tissue
transfers.

Glass

Bullseye Glass Co.
USA
www.bullseyeglass.com
Supplier of a broad range of compatible
art glass. Website has very useful 'tip
sheets'.

Pearsons Glass
www.pearsonsglass.com
Supplier of stained and decorative glass
and accessories. Useful tutorials on
website.

Further reading

Adam, R. & Robertson, C., *Screenprinting: The complete water-based system* (London: Thames and Hudson, 2005).

Darty, L., *The Art of Enameling: Techniques, projects, and inspiration* (NY: Lark Books, 2004).

Fortune, D. *The Art Teacher's Guide to Water-based Screenprinting* (Daler Rowney, 2006).

Gale, C., *Etching and Photopolymer Intaglio Techniques* (London: A&C Black, 2006).

Hoskins, S., *Water-based Screenprinting* (London: A&C Black, 2003).

Petrie, K., *Glass and Print* (London: A&C Black, 2006).

Petrie, K., 'Mix, Match, and Madness: Early transfer printing and its potential for the contemporary artist' in *Ceramics Technical* 4, May 1997, 17–25.

Scott, P., *Ceramics and Print* (London: A&C Black, 1994).

Turrell, E. (ed.), *Contemporary Print in Enamel* (Bristol: Impact Press, 2001).

Turrell, E. (ed.), *The Enamel Experience: International Badge Exhibition* (Bristol: Impact Press, 2007).

Wandless, P.A., *Image Transfer on Clay* (NY: Lark Books, 2006).

Also see endnotes for additional reading.

Index